Student Edition

Eureka Math

Grade 1
Module 2

Special thanks go to the Gordon A. Cain Center and to the Department of Mathematics at Louisiana State University for their support in the development of *Eureka Math*.

For a free *Eureka Math* Teacher Resource Pack, Parent Tip Sheets, and more please visit www.Eureka.tools

Published by the non-profit Great Minds

Printed in the U.S.A.
This book may be purchased from the publisher at eureka-math.org
10 9 8

ISBN 978-1-63255-289-1

Name GAbriel Date 2020

Read the math story. Make a simple math drawing with labels. Circle 10 and solve.

1. Bill went to the store. He bought 1 apple, 9 bananas, and 6 pears. How many pieces of fruit did he buy in all?

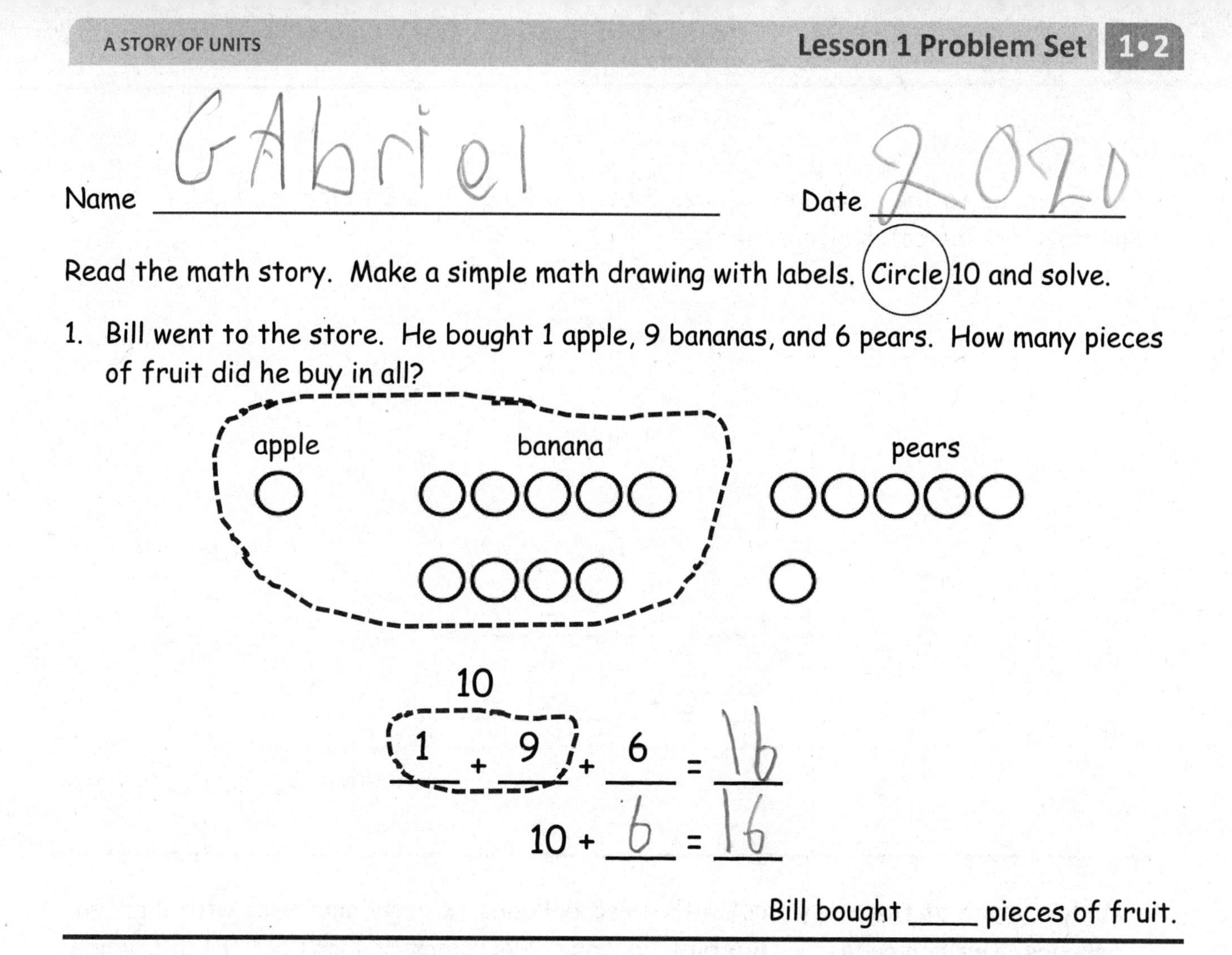

10

1 + 9 + 6 = 16

10 + 6 = 16

Bill bought ____ pieces of fruit.

2. Maria gets some new toys for her birthday. She gets 4 dolls, 7 balls, and 3 games. How many toys did she receive?

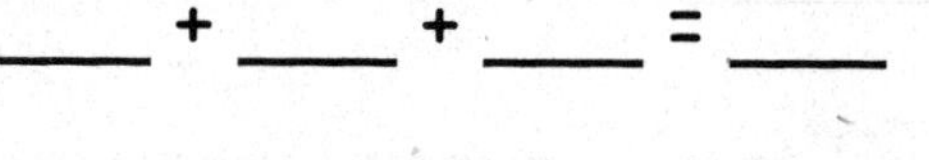

____ + ____ + ____ = ____

10 + ____ = ____

Maria received ____ toys.

3. Maddy goes to the pond and catches 8 bugs, 3 frogs, and 2 tadpoles. How many animals did she catch altogether?

___ + ___ + ___ = ___

10 + ___ = ___

Maddy caught ____ animals.

---

4. Molly arrived at the party first with 4 red balloons. Kenny came next with 2 green balloons. Dara came last with 6 blue balloons. How many balloons did these friends bring?

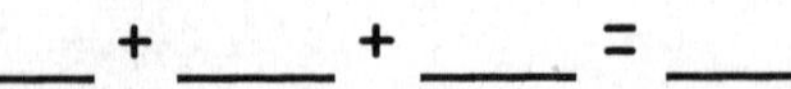

There are ____ balloons.

Name ______________________________ Date ______________

Read the math story. Make a simple math drawing with labels. Circle 10 and solve.

1. Chris bought some treats. He bought 5 granola bars, 6 boxes of raisins, and 4 cookies. How many treats did Chris buy?

___ + ___ + ___ = ___

10 + ___ = ___

Chris bought ____ treats.

---

2. Cindy has 5 cats, 7 goldfish, and 5 dogs. How many pets does she have in all?

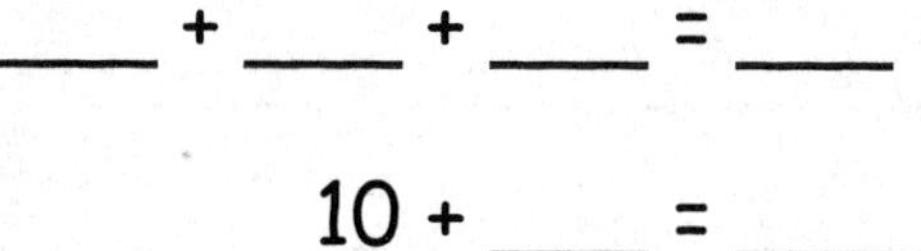

___ + ___ + ___ = ___

10 + ___ = ___

Cindy has ____ pets.

3. Mary gets stickers at school for good work. She got 7 puffy stickers, 6 smelly stickers, and 3 flat stickers. How many stickers did Mary get at school altogether?

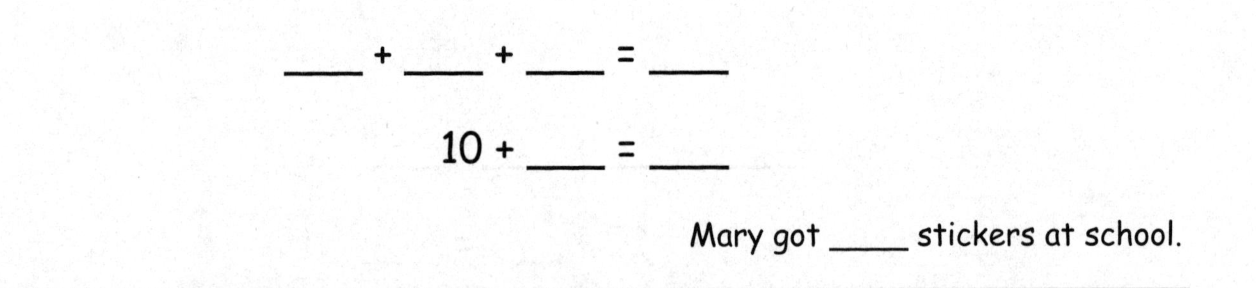

____ + ____ + ____ = ____

10 + ____ = ____

Mary got ____ stickers at school.

---

4. Jim sat at a table with 4 teachers and 9 children. How many people were at the table after Jim sat down?

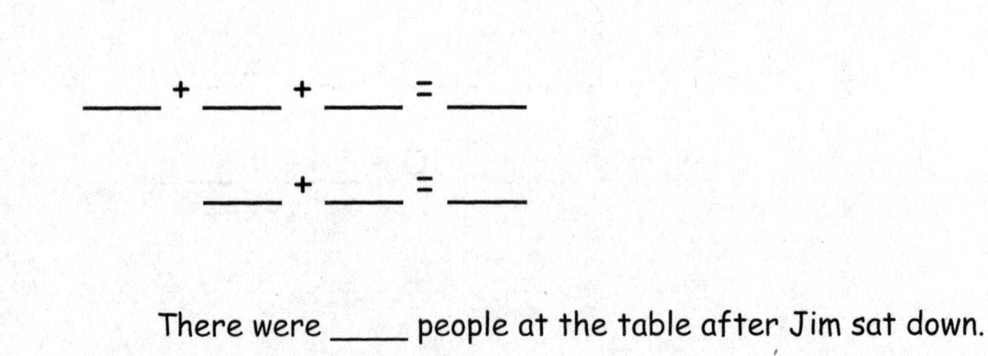

____ + ____ + ____ = ____

____ + ____ = ____

There were ____ people at the table after Jim sat down.

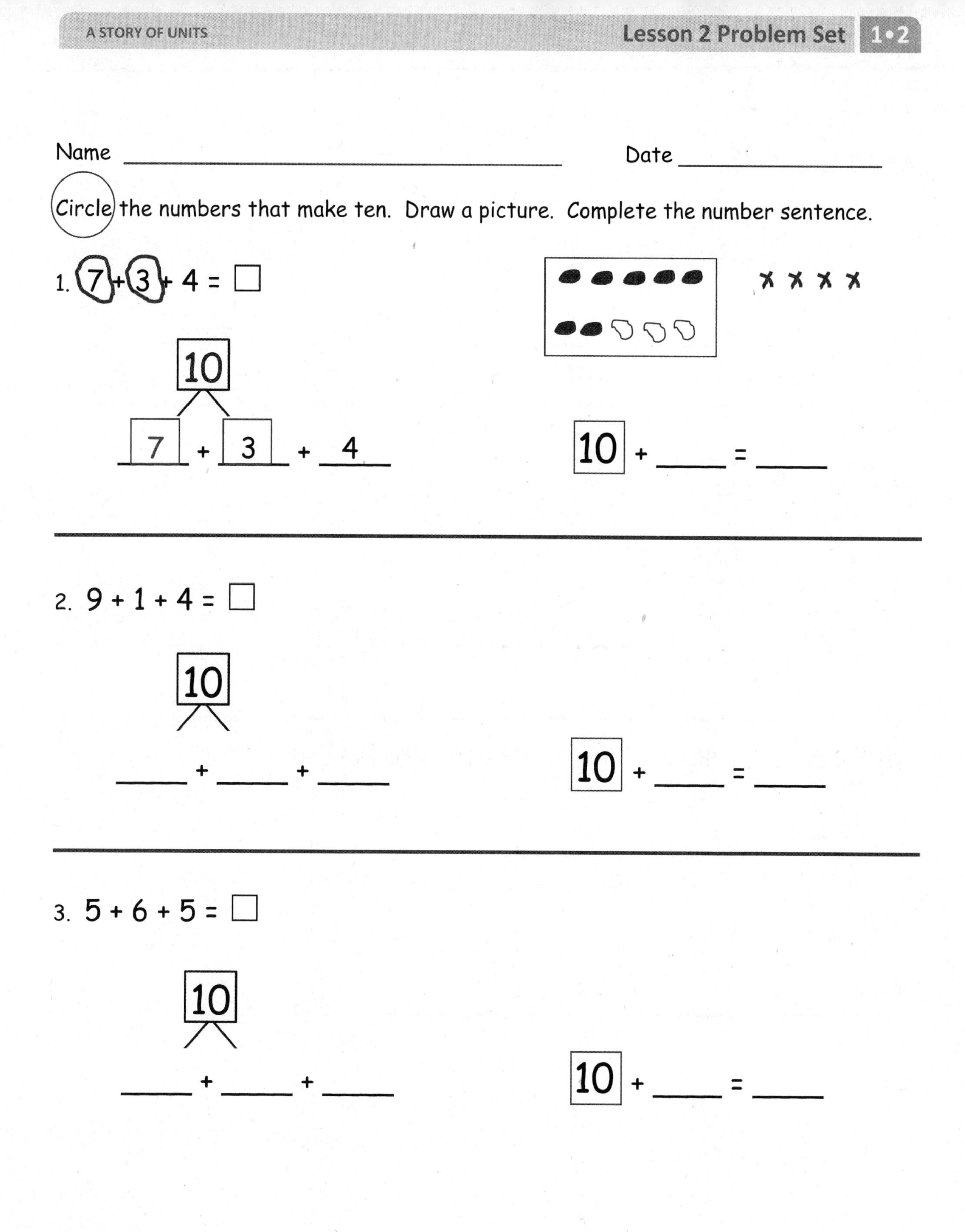

Name ______________________ Date ____________

Circle the numbers that make ten. Draw a picture. Complete the number sentence.

1. 7 + 3 + 4 = □

10

7 + 3 + 4

10 + ____ = ____

2. 9 + 1 + 4 = □

10

____ + ____ + ____

10 + ____ = ____

3. 5 + 6 + 5 = □

10

____ + ____ + ____

10 + ____ = ____

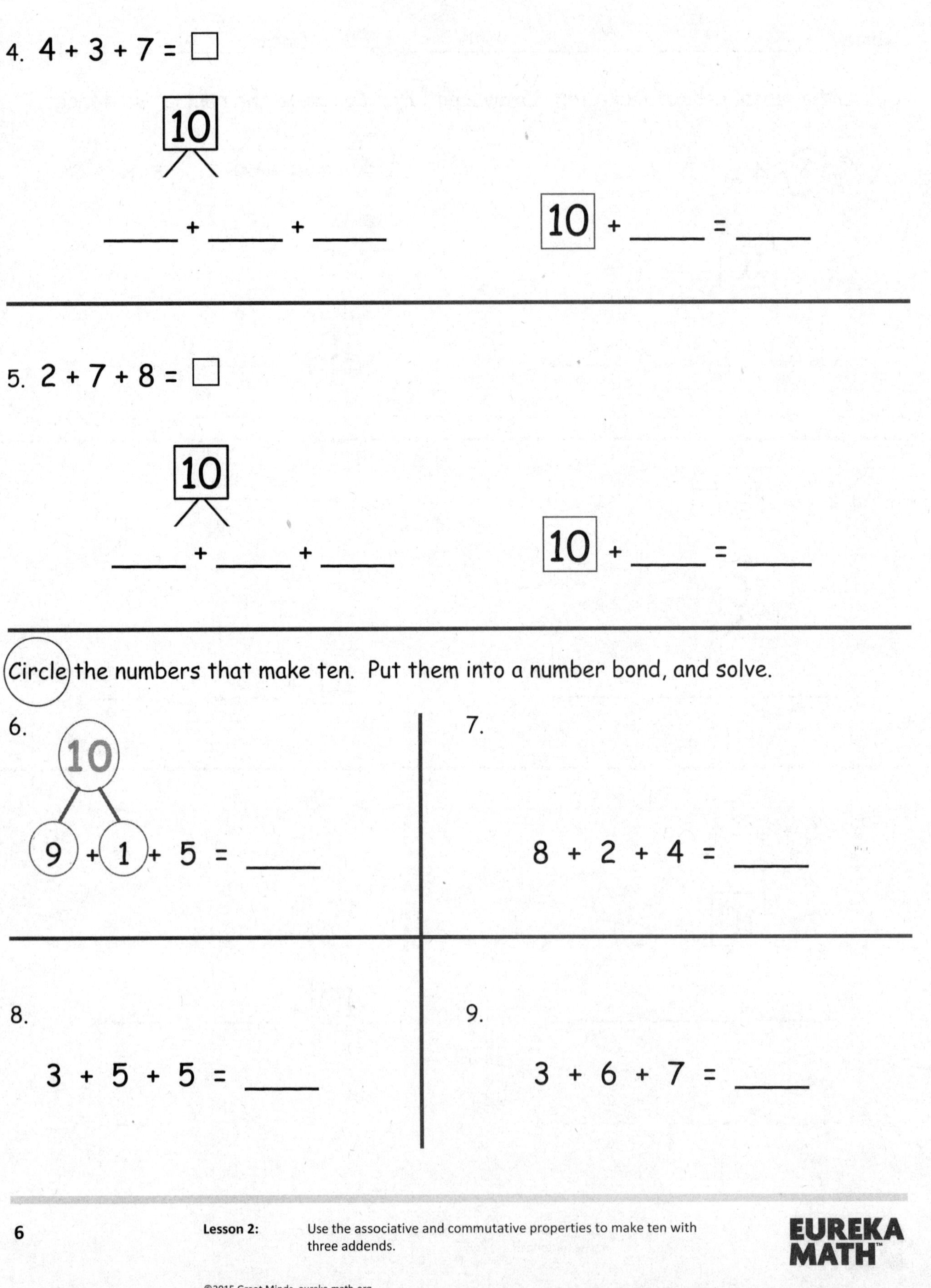

4. 4 + 3 + 7 = ☐

10

___ + ___ + ___

10 + ___ = ___

5. 2 + 7 + 8 = ☐

10

___ + ___ + ___

10 + ___ = ___

Circle the numbers that make ten. Put them into a number bond, and solve.

6.

10

9 + 1 + 5 = ___

7.

8 + 2 + 4 = ___

8.

3 + 5 + 5 = ___

9.

3 + 6 + 7 = ___

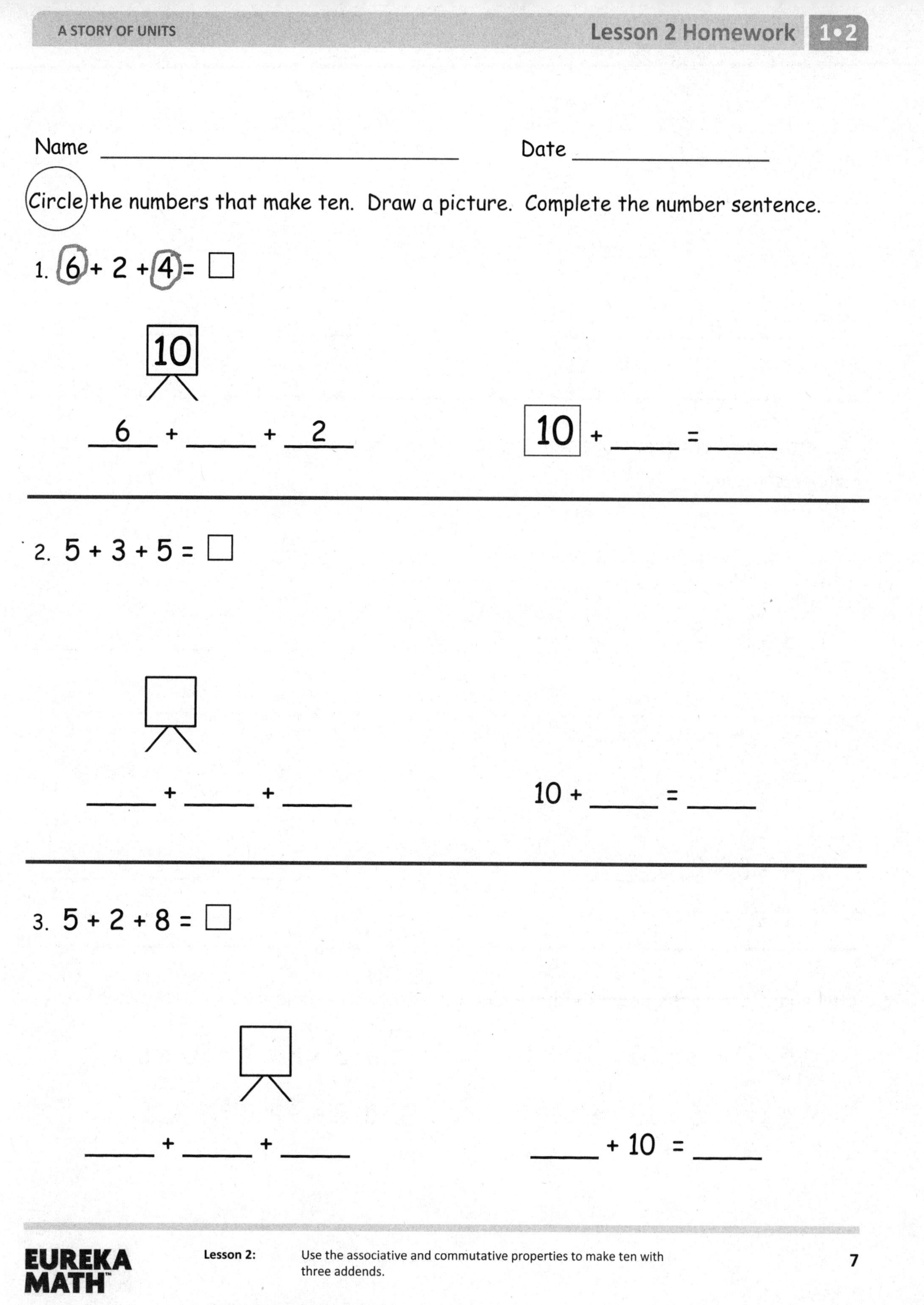

Name ______________________________ Date ________________

Circle the numbers that make ten. Draw a picture. Complete the number sentence.

1. 6 + 2 + 4 = ☐

10

_6_ + ____ + _2_ 10 + ____ = ____

2. 5 + 3 + 5 = ☐

____ + ____ + ____ 10 + ____ = ____

3. 5 + 2 + 8 = ☐

____ + ____ + ____ ____ + 10 = ____

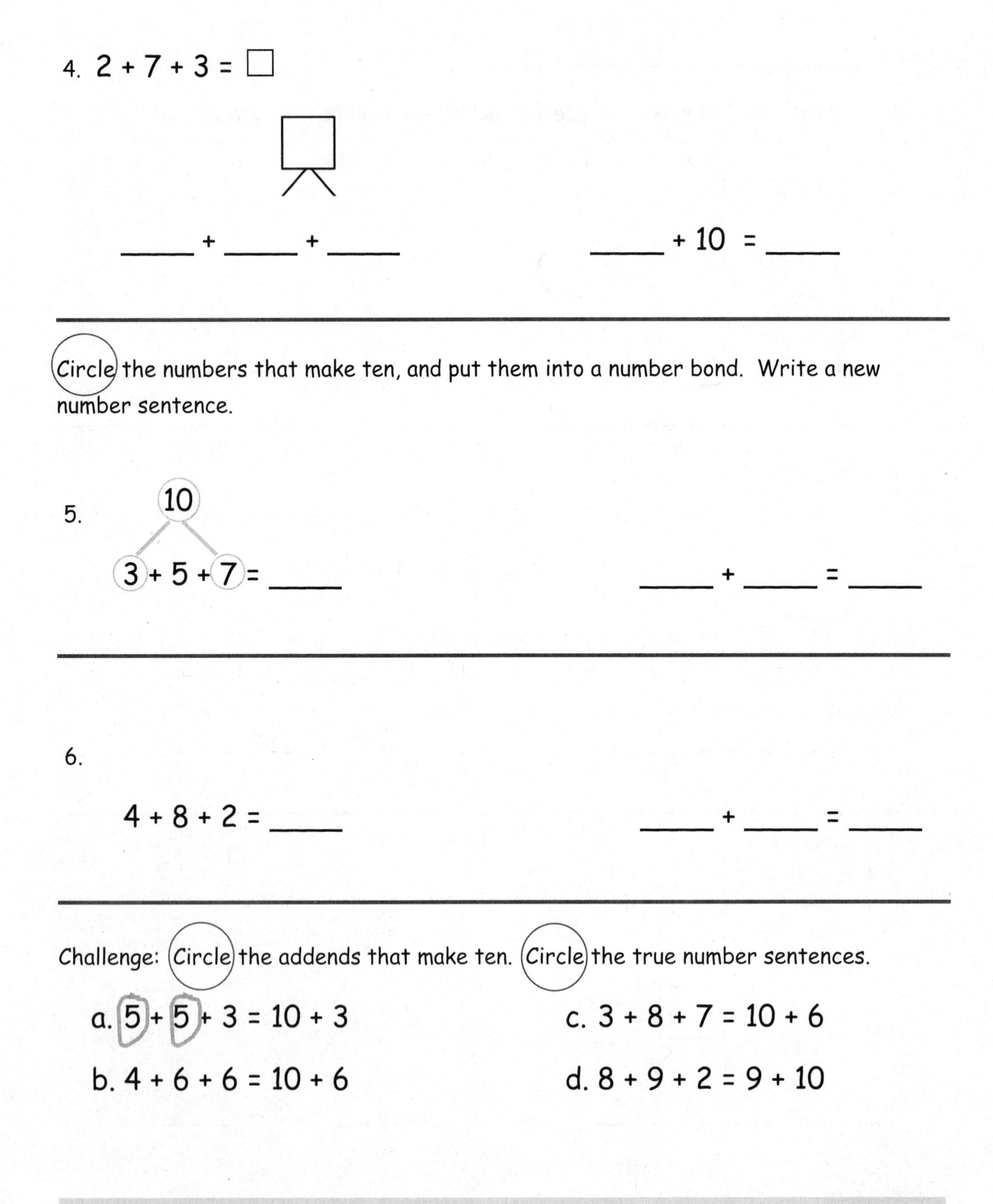

4. 2 + 7 + 3 = ☐

_____ + _____ + _____ _____ + 10 = _____

Circle the numbers that make ten, and put them into a number bond. Write a new number sentence.

5. 10

3 + 5 + 7 = _____ _____ + _____ = _____

6.

4 + 8 + 2 = _____ _____ + _____ = _____

Challenge: Circle the addends that make ten. Circle the true number sentences.

a. 5 + 5 + 3 = 10 + 3

b. 4 + 6 + 6 = 10 + 6

c. 3 + 8 + 7 = 10 + 6

d. 8 + 9 + 2 = 9 + 10

Name ______________________________ Date ______________

Draw and circle to show how you made ten to help you solve the problem.

1. Maria has 9 snowballs, and Tony has 6. How many snowballs do they have in all?

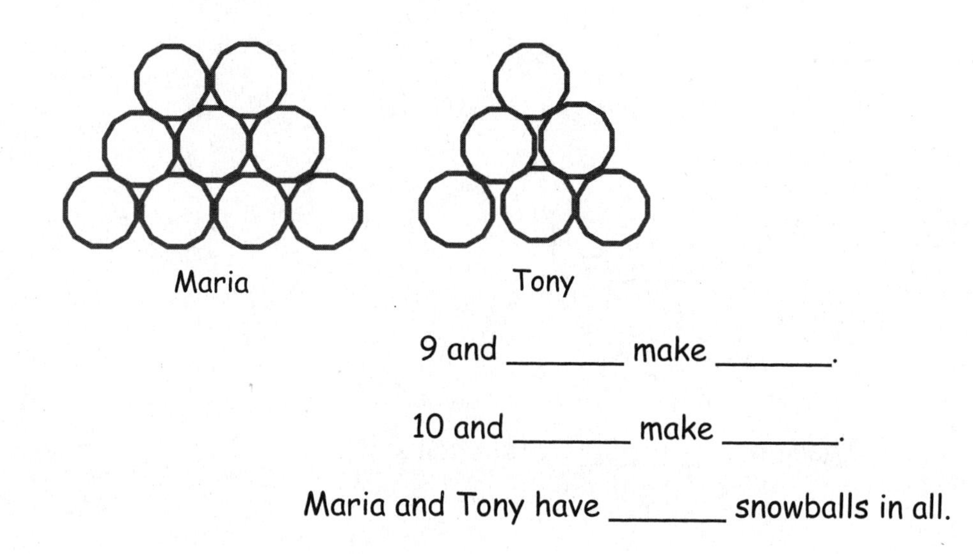

9 and ______ make ______.

10 and ______ make ______.

Maria and Tony have ______ snowballs in all.

2. Bob has 9 raisins, and Jonny has 4. How many raisins do they have altogether?

9 + ___ = ___

10 + ___ = ___

Bob and Jonny have ______ raisins altogether.

3. There are 3 chairs on the left side of the classroom and 9 on the right side. How many total chairs are in the classroom?

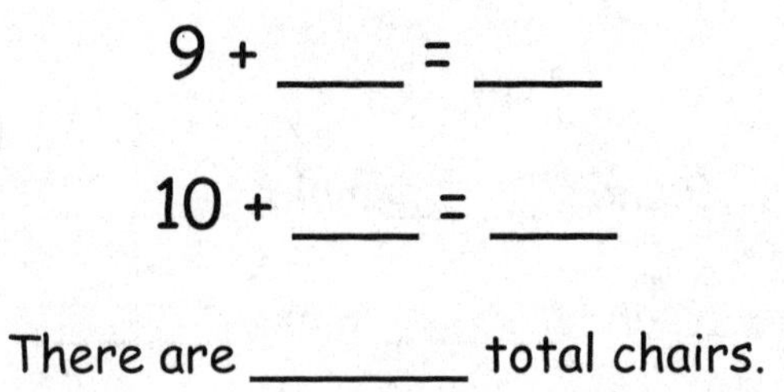

9 + ____ = ____

10 + ____ = ____

There are ______ total chairs.

4. There are 7 children sitting on the rug and 9 children standing. How many children are there in all?

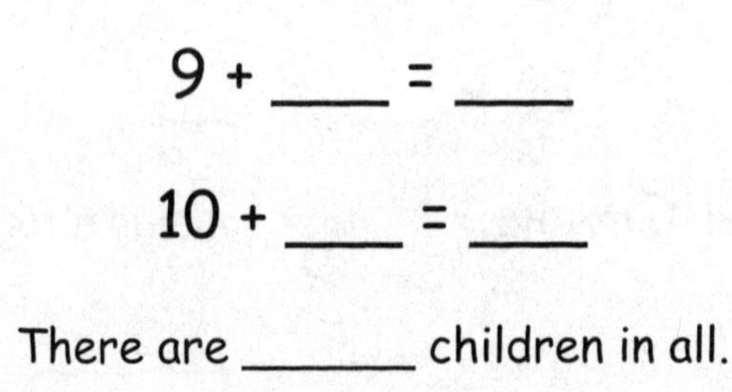

9 + ____ = ____

10 + ____ = ____

There are ______ children in all.

Name ______________________________ Date ______________

Draw, label, and circle to show how you made ten to help you solve.
Complete the number sentences.

1. Ron has 9 marbles, and Sue has 4 marbles.
   How many marbles do they have in all?

9 and ______ make ______.

10 and ______ make ______.

Ron and Sue have _____ marbles.

---

2. Jim has 5 cars, and Tina has 9. How many cars do they have altogether?

9 and ______ make ______.

10 and ______ make ______.

Jim and Tina have ___ cars.

3. Stan has 6 fish, and Meg has 9. How many fish do they have in all?

9 + 1 = 10

10 + 6 = 16

Stan and Meg have 16 fish.

4. Rick made 7 cookies, and Mom made 9. How many cookies did Rick and Mom make?

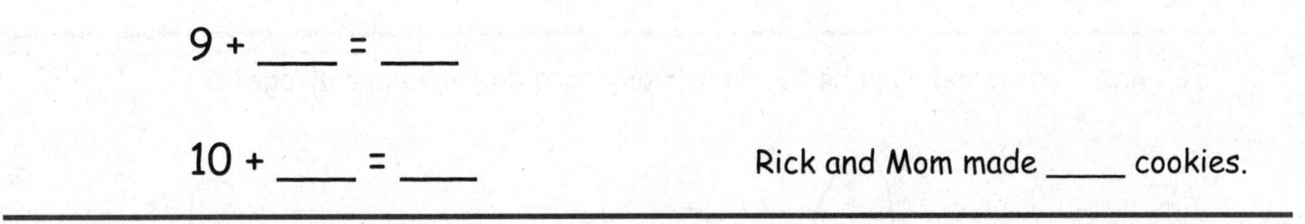

9 + ___ = ___

10 + ___ = ___

Rick and Mom made ____ cookies.

5. Dad has 8 pens, and Tony has 9. How many pens do Dad and Tony have in all?

9 + ___ = ___

10 + ___ = ___

Dad and Tony have ___ pens.

G1-M2-SE-B2-1.3.1-12.2015

Name Gabriel Date 2020

Change the picture to make ten. Write the easier number sentence and solve.

1. Tom has 9 red pencils and 5 yellow. How many pencils does Tom have in all?

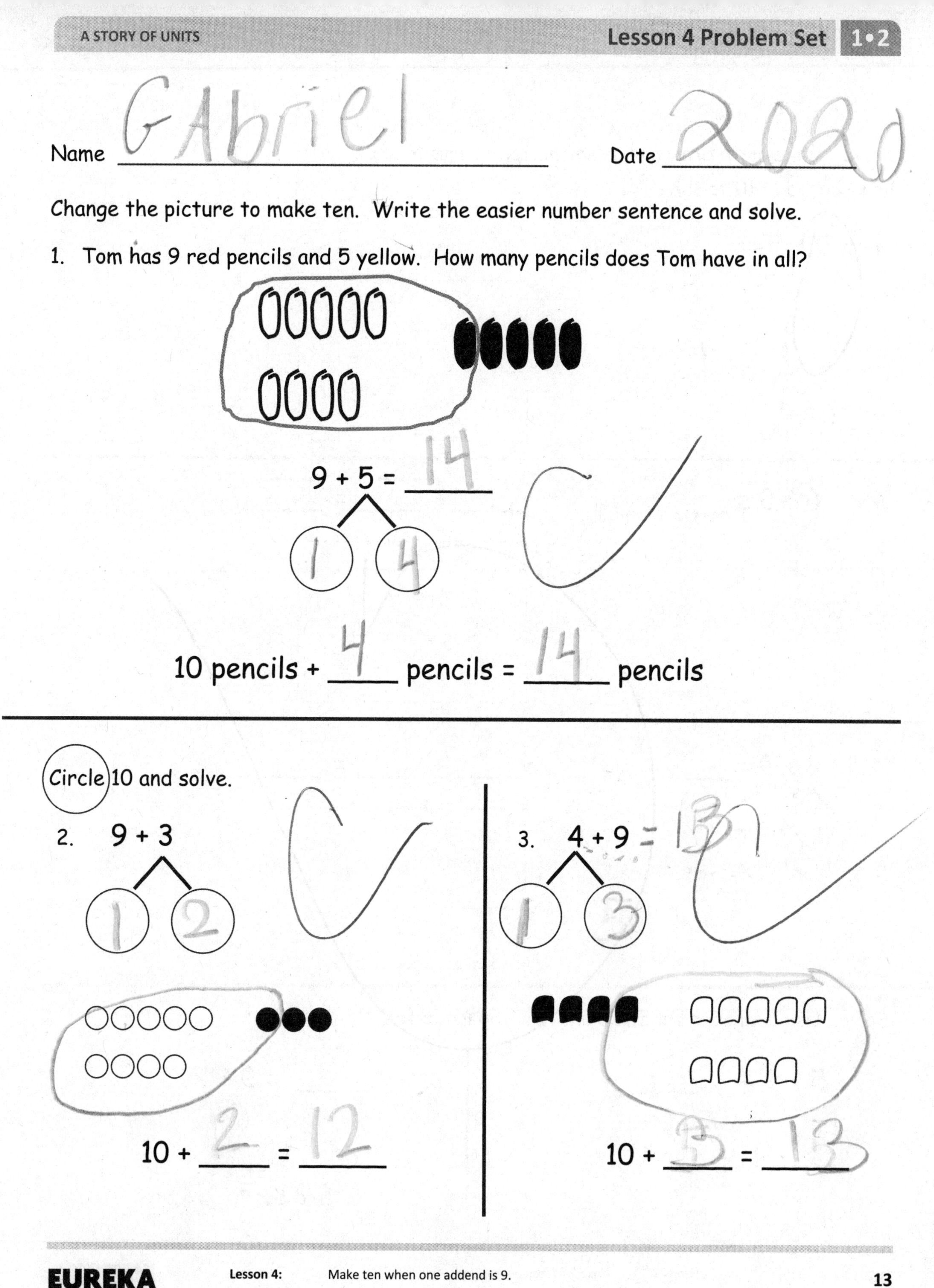

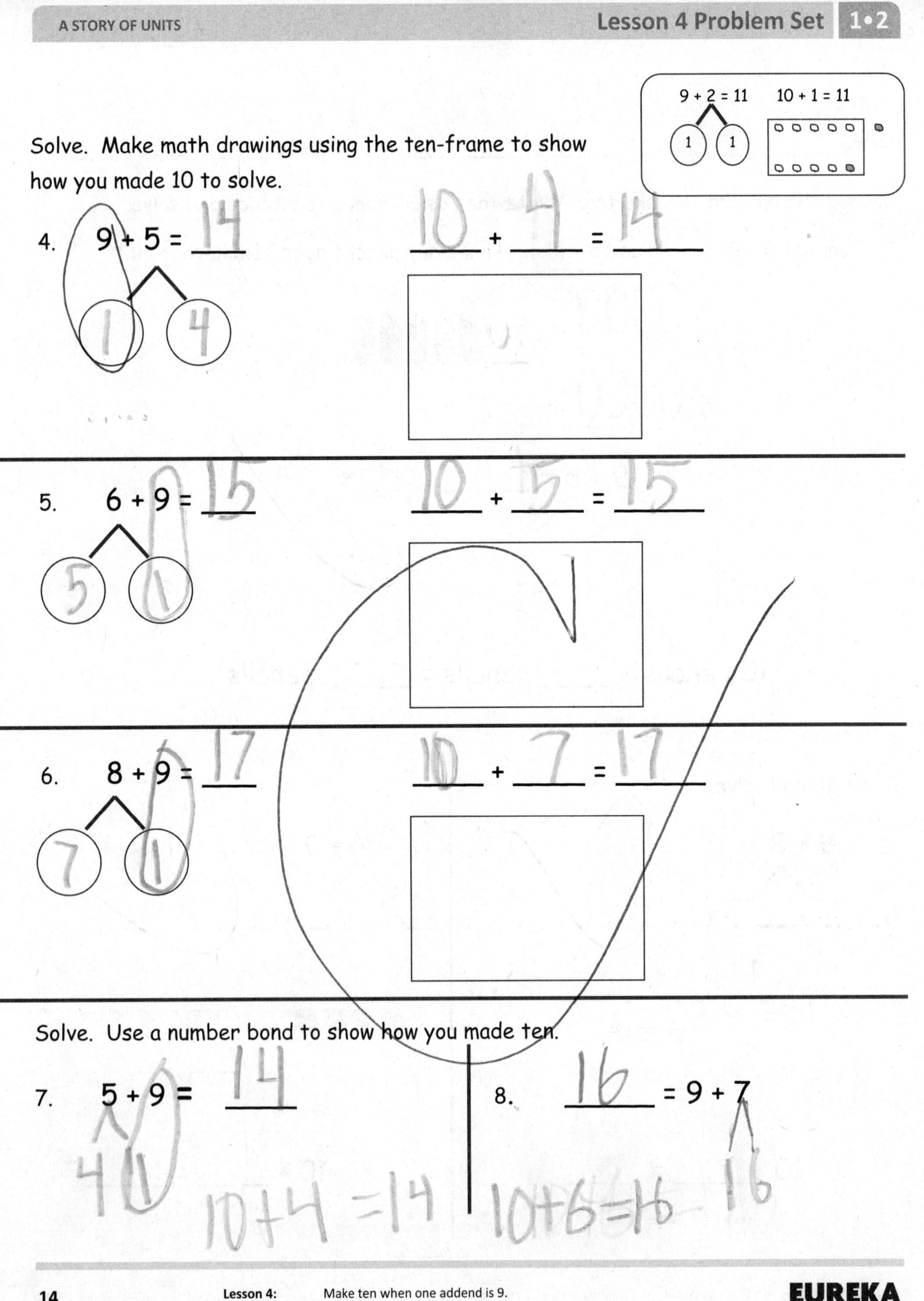

Solve. Make math drawings using the ten-frame to show how you made 10 to solve.

9 + 2 = 11   10 + 1 = 11

4. 9 + 5 = ____   ____ + ____ = ____

5. 6 + 9 = ____   ____ + ____ = ____

6. 8 + 9 = ____   ____ + ____ = ____

Solve. Use a number bond to show how you made ten.

7. 5 + 9 = ____

8. ____ = 9 + 7

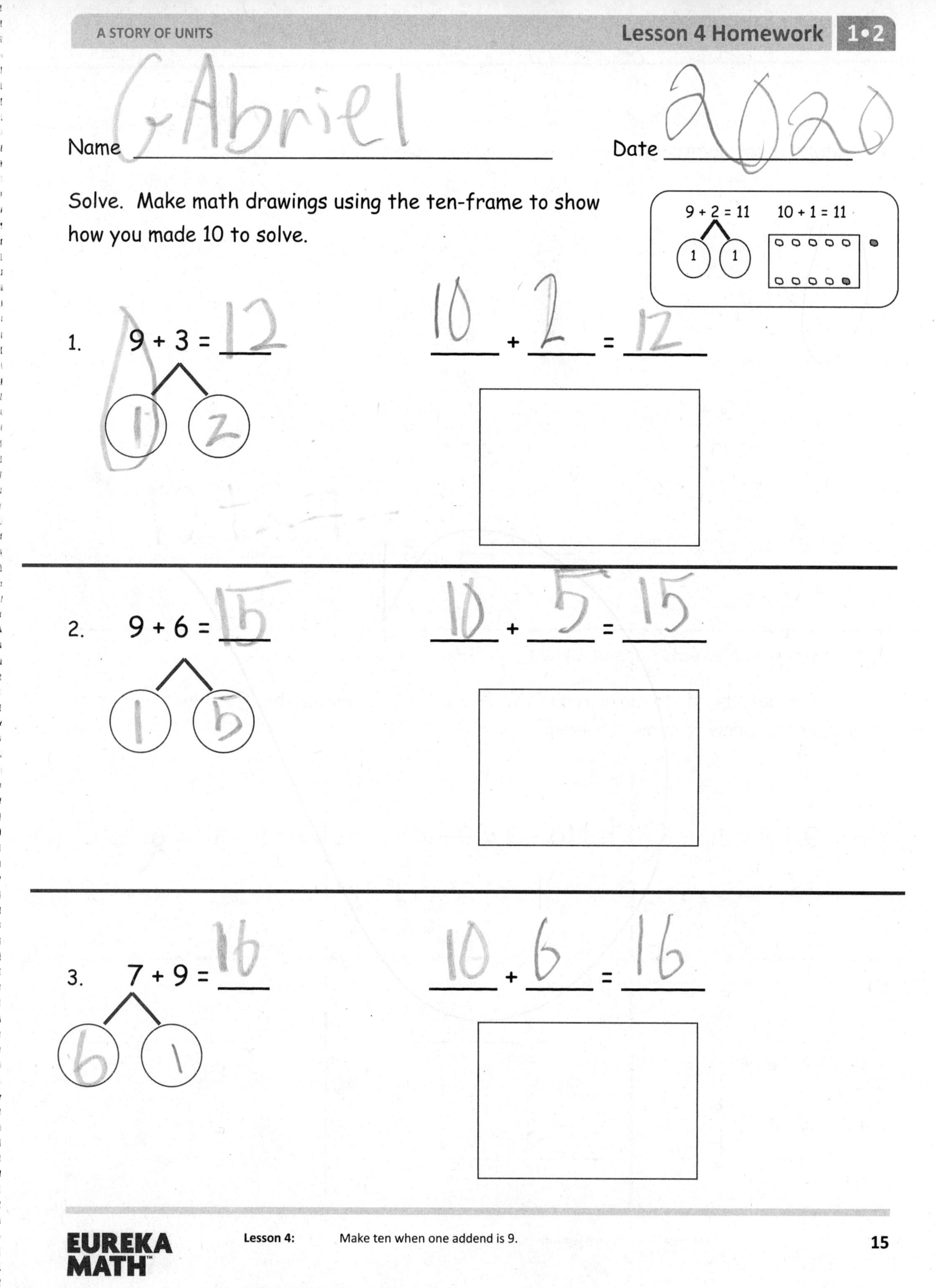

Name Gabriel Date 2020

Solve. Make math drawings using the ten-frame to show how you made 10 to solve.

9 + 2 = 11 10 + 1 = 11

1. 9 + 3 = 12 10 + 2 = 12

2. 9 + 6 = 15 10 + 5 = 15

3. 7 + 9 = 16 10 + 6 = 16

4. Match the number sentences to the bonds you used to help you make ten.

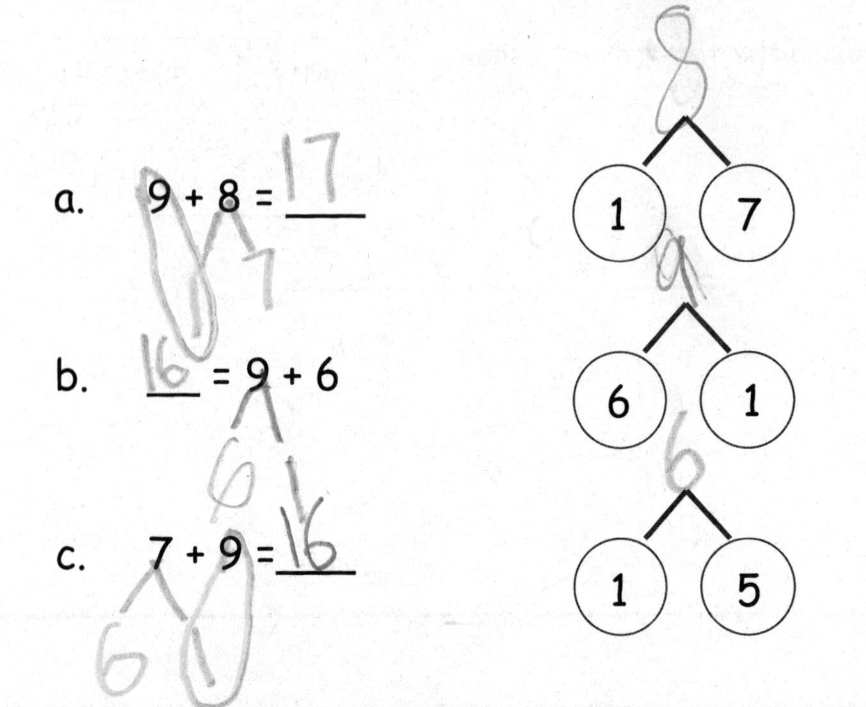

5. Show how the expressions are equal.

Use numbers bonds to make ten in the 9+ *fact* expression within the true number sentence. Draw to show the total.

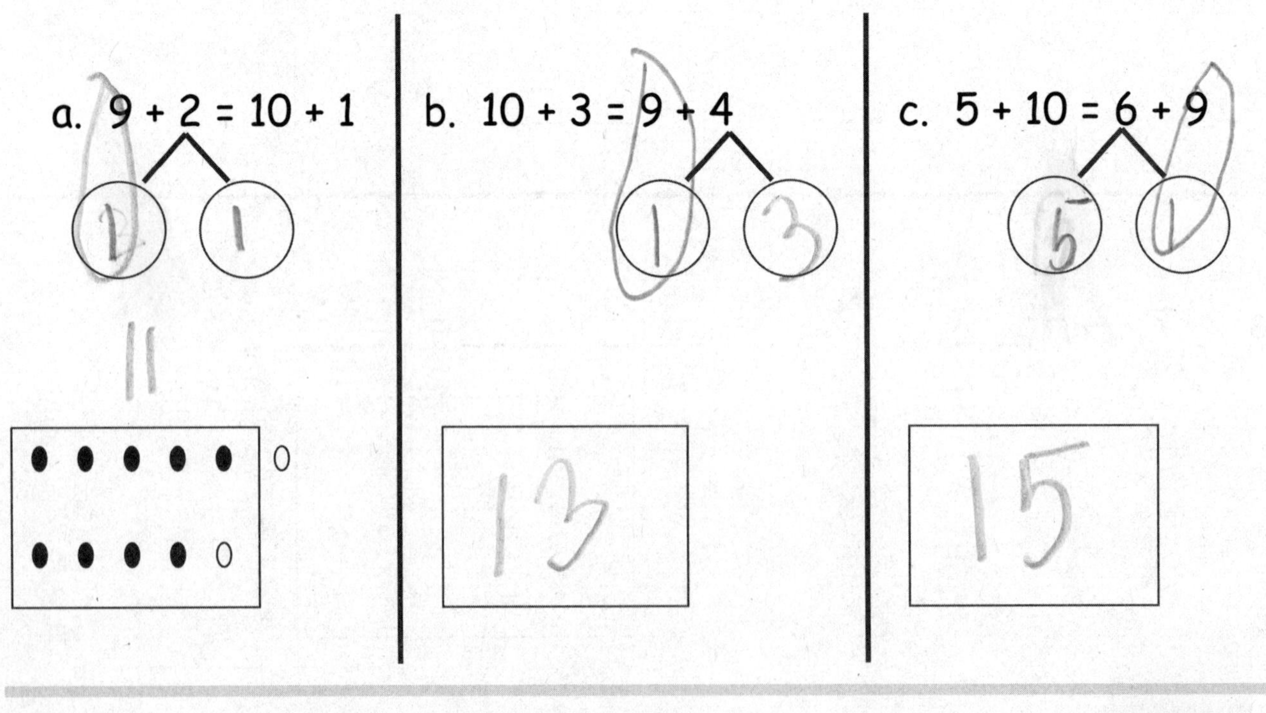

Name ______________________________ Date ________________

Make ten to solve. Use the number bond to show how you took the 1 out.

1. Sue has 9 tennis balls and 3 soccer balls. How many balls does she have?

9 + 3 = ____ 10 + ___ = ___

Sue has _____ balls.

2. 9 + 4 = ____ 10 + ___ = ___

Use number bonds to show your thinking. Write the 10+ fact.

3. 9 + 2 = _____ _____ + _____ = _____

4. 9 + 5 = _____ _____ + _____ = _____

5. 9 + 4 = _____ _____ + _____ = _____

6. 9 + 7 = ____ ____ + ____ = ____

7. 9 + ____ = ____ 10 + 7 = ____

Complete the addition sentences.

8. a. 10 + 1 = ____ 11

b. 9 + 2 = ____ 11

9. a. 10 + 8 = ____ 18

b. 9 + 9 = ____ 18

10. a. 10 + 7 = ____

b. 9 + 8 = ____

11. a. 5 + 10 = ____

b. 6 + 9 = ____

12. a. 6 + 10 = ____

b. 7 + 9 = ____

Name ________________________________ Date ______________

Solve the number sentences. Use number bonds to show your thinking. Write the 10+ fact and new number bond.

1. 9 + 6 = _____ 10 + _____ = _____

2. 9 + 8 = _____ _____ + _____ = _____

3. 5 + 9 = _____ _____ + _____ = _____

4. 7 + 9 = _____ _____ + _____ = _____

5. Solve. Match the number sentence to the 10+ number bond.

a. 9 + 5 = _____ b. 9 + 6 = _____ c. 9 + 8 = _____

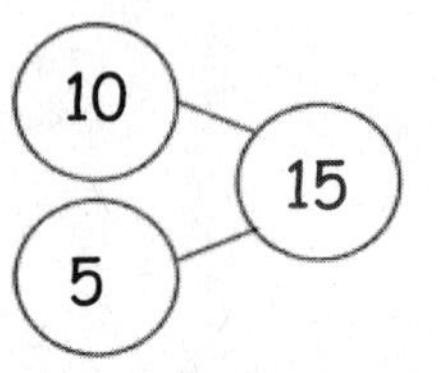

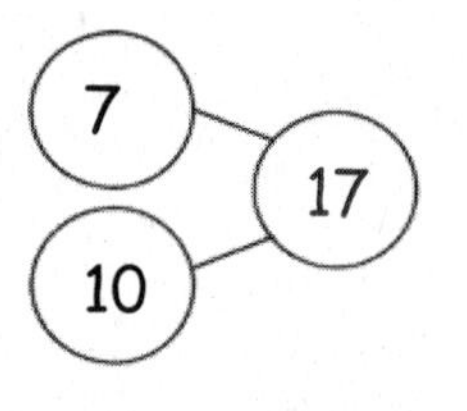

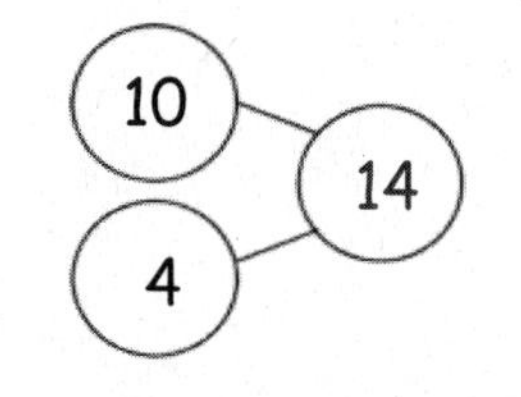

Use an efficient strategy to solve the number sentences.

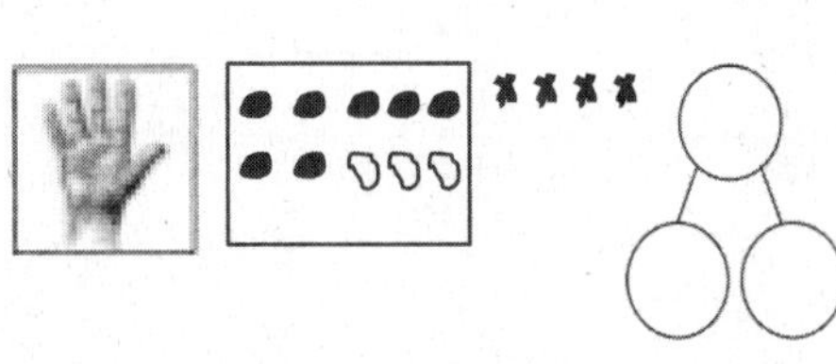

6. 9 + 7 = _____ 7. 9 + 2 = _____ 8. 9 + 1 = _____

9. 8 + 9 = _____ 10. 4 + 9 = _____ 11. 9 + 9 = _____

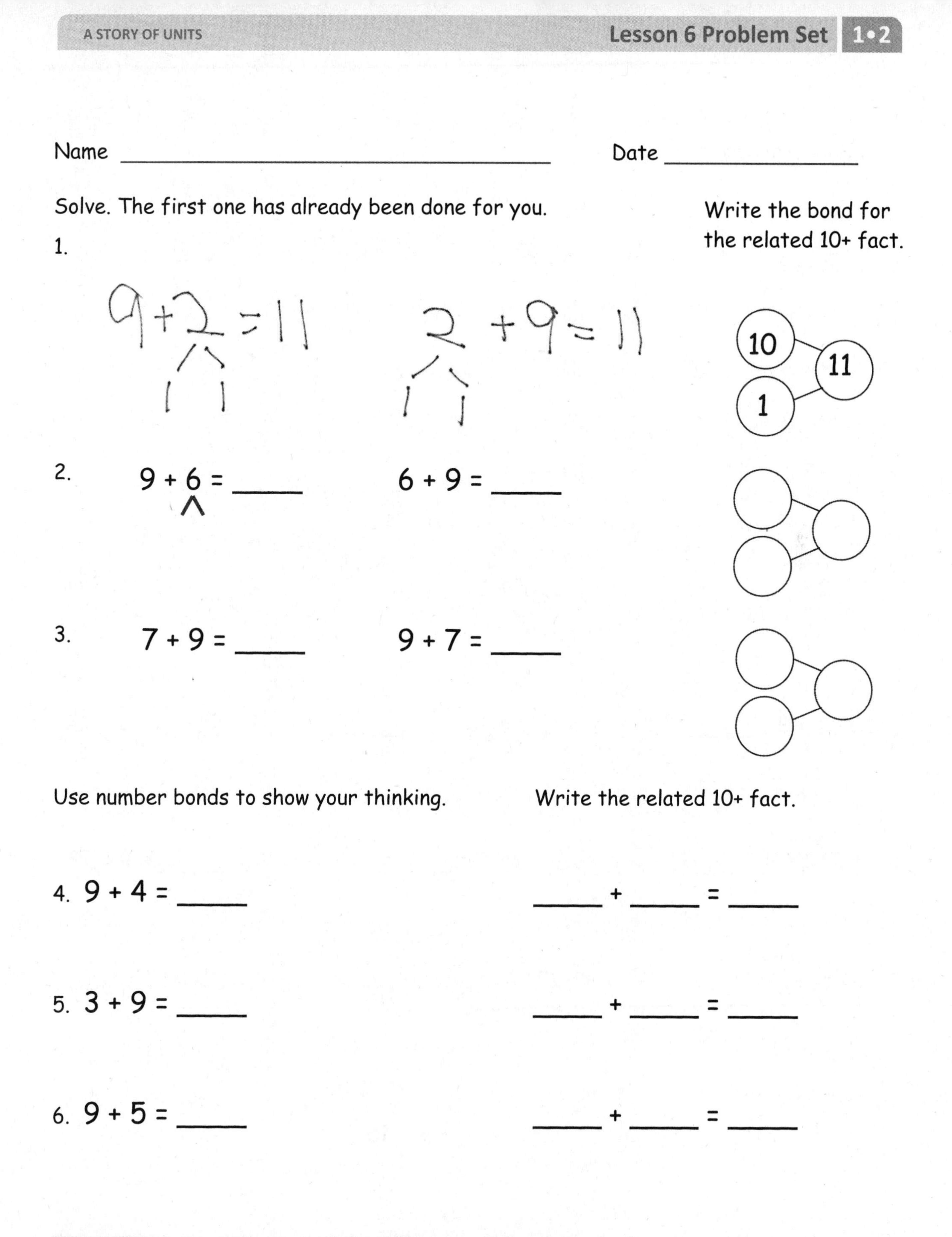

Name ______________________ Date ____________

Solve. The first one has already been done for you.

Write the bond for the related 10+ fact.

1. 9 + 2 = 11 2 + 9 = 11

2. 9 + 6 = _____ 6 + 9 = _____

3. 7 + 9 = _____ 9 + 7 = _____

Use number bonds to show your thinking.

Write the related 10+ fact.

4. 9 + 4 = _____ _____ + _____ = _____

5. 3 + 9 = _____ _____ + _____ = _____

6. 9 + 5 = _____ _____ + _____ = _____

7. Match the equal expressions.

a. 9 + 3	10 + 4
b. 5 + 9	10 + 0
c. 9 + 6	10 + 2
d. 8 + 9	10 + 5
e. 9 + 7	10 + 7
f. 9 + 1	10 + 6

8. Complete the addition sentences to make them true.

a. 2 + 10 = _____

b. 7 + 9 = _____

c. _____ + 10 = 14

d. 3 + 9 = _____

e. 3 + 10 = _____

f. _____ + 9 = 14

g. 10 + 9 = _____

h. 8 + 9 = _____

i. _____ + 7 = 17

j. 5 + 9 = _____

k. ____ + 10 = 18

l. _____ + 9 = 17

m. 6 + 10 = _____

n. _____ + 9 = 16

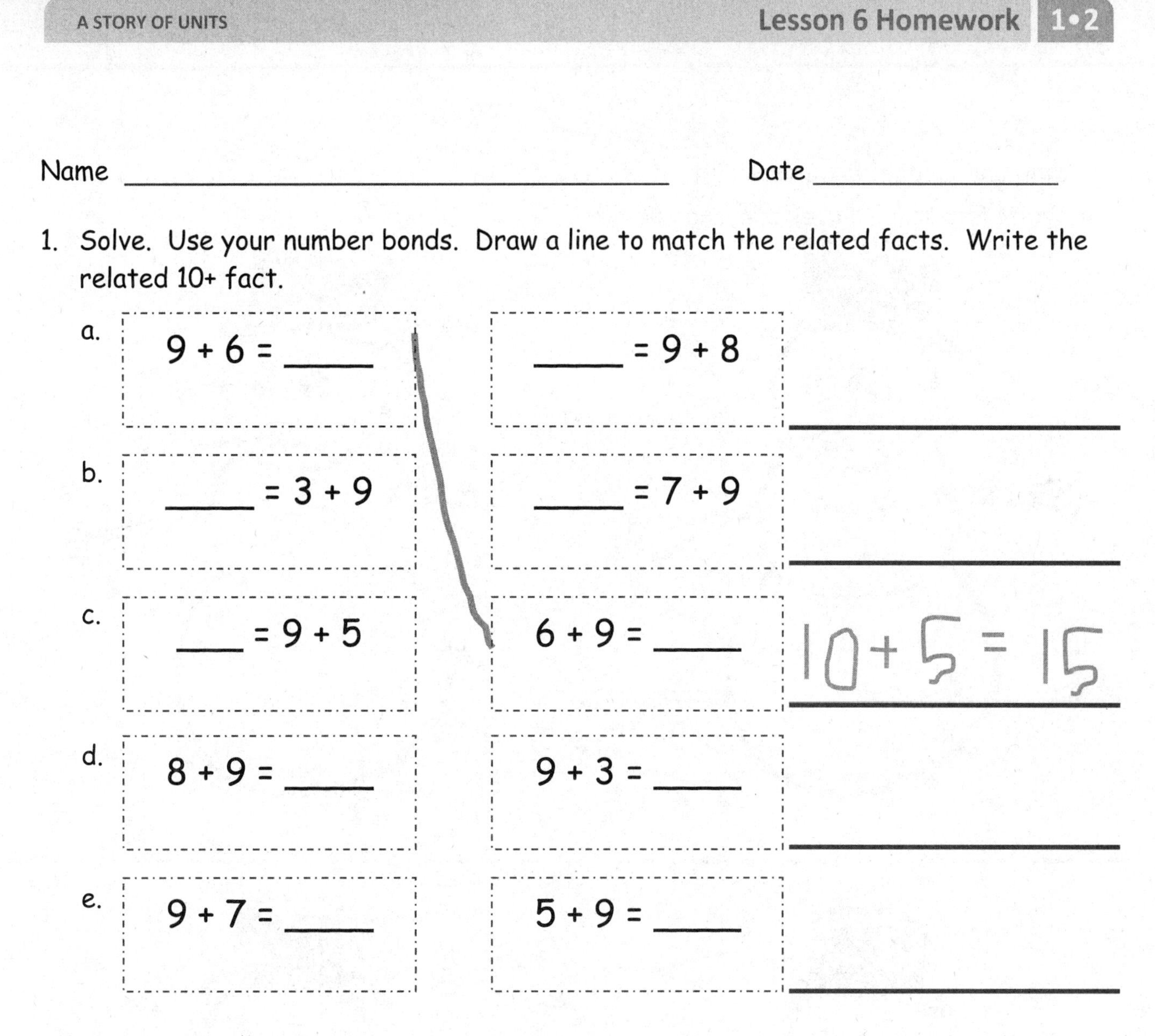

Name ______________________ Date __________

1. Solve. Use your number bonds. Draw a line to match the related facts. Write the related 10+ fact.

a. $9 + 6 =$ ____ ____ $= 9 + 8$ ____________

b. ____ $= 3 + 9$ ____ $= 7 + 9$ ____________

c. ____ $= 9 + 5$ $6 + 9 =$ ____ 10 + 5 = 15

d. $8 + 9 =$ ____ $9 + 3 =$ ____ ____________

e. $9 + 7 =$ ____ $5 + 9 =$ ____ ____________

2. Complete the addition sentences to make them true.

a. $3 + 10 =$ ____

b. $4 + 9 =$ ____

c. $10 + 5 =$ ____

d. $9 + 6 =$ ____

e. $7 + 10 =$ ____

f. ____ $= 7 + 9$

g. $10 +$ ____ $= 18$

h. $9 + 8 =$ ____

i. ____ $+ 9 = 19$

j. $5 + 9 =$ ____

3. Find and color the expression that is equal to the expression on the snowman's hat.
   Write the true number sentence below.

Name ______________________________ Date ______________

Circle to show how you made ten to help you solve.

1. John has 8 tennis balls. Toni has 5. How many tennis balls do they have in all?

John

Toni

8 and ______ make ______.

10 and ______ make ______.

John and Toni have ______ tennis balls in all.

2. Bob has 8 raisins, and Jenny has 4. How many raisins do they have altogether?

8 and ______ make ______.

10 and ______ make ______.

Bob and Jenny have ______ raisins altogether.

3. There are 3 chairs on the right side of the classroom and 8 on the left side. How many total chairs are in the classroom?

8 and ______ make ______.

10 and ______ make ______.

There are _______ total chairs.

4. There are 7 children sitting on the rug and 8 children standing. How many children are there in all?

8 and ______ make ______.

10 and ______ make ______.

There are ______ children in all.

Name ______________________________ Date ______________

Draw, label, and circle to show how you made ten to help you solve.

Write the number sentences you used to solve.

00000000 ●●●
W B

1. Meg gets 8 toy animals and 4 toy cars at a party.
   How many toys does Meg get in all?

8 + 3 = 11
10 + 1 = 11

8 + 4 = _____

10 + _____ = _____ Meg gets _____ toys.

2. John makes 6 baskets in his first basketball game and 8 baskets in his second.
   How many baskets does he make altogether?

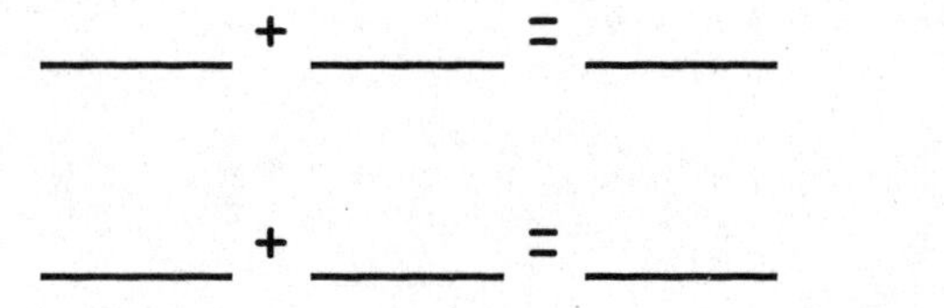

_____ + _____ = _____

_____ + _____ = _____ John makes _____ baskets.

3. May has a party. She invites 7 girls and 8 boys. How many friends does she invite in all?

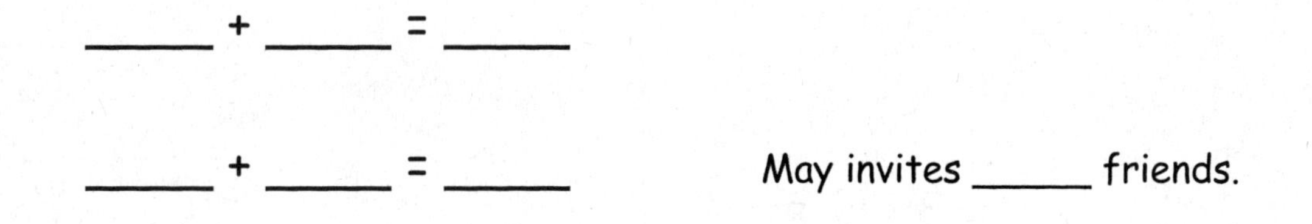

_____ + _____ = _____

_____ + _____ = _____ May invites _____ friends.

4. Alec collects baseball hats. He has 9 Mets hats and 8 Yankees hats. How many hats are in his collection?

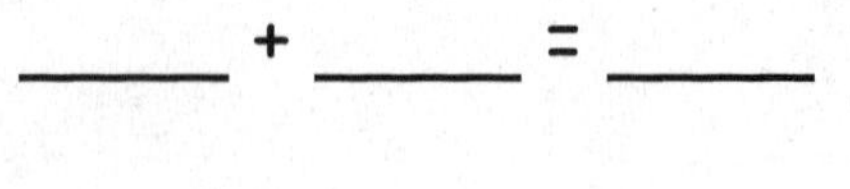

_____ + _____ = _____

_____ + _____ = _____ Alec has _____ hats.

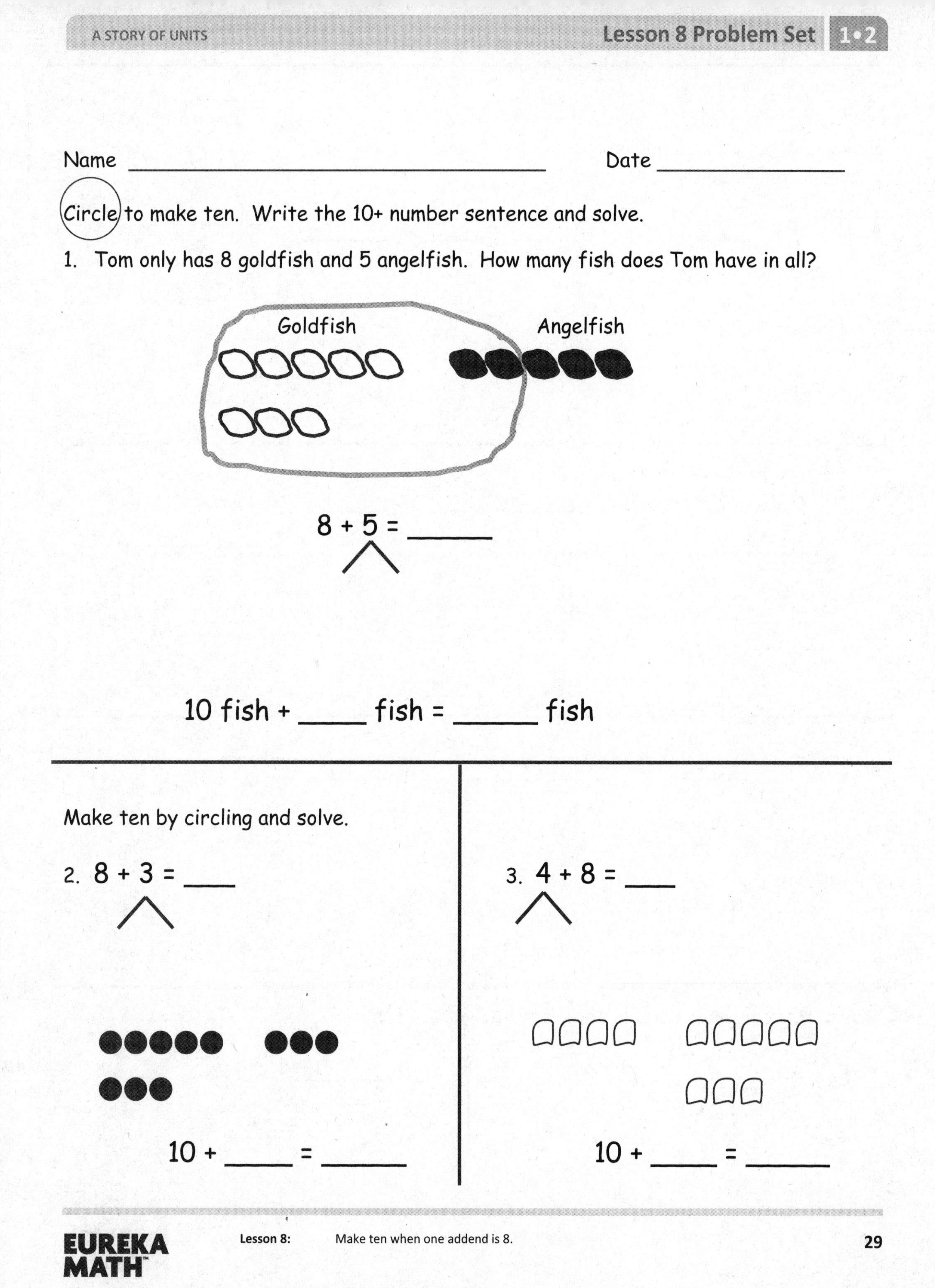

Name ______________________________ Date ______________

Circle to make ten. Write the 10+ number sentence and solve.

1. Tom only has 8 goldfish and 5 angelfish. How many fish does Tom have in all?

8 + 5 = _____

10 fish + _____ fish = _____ fish

Make ten by circling and solve.

2. 8 + 3 = ____

10 + _____ = _____

3. 4 + 8 = ____

10 + _____ = _____

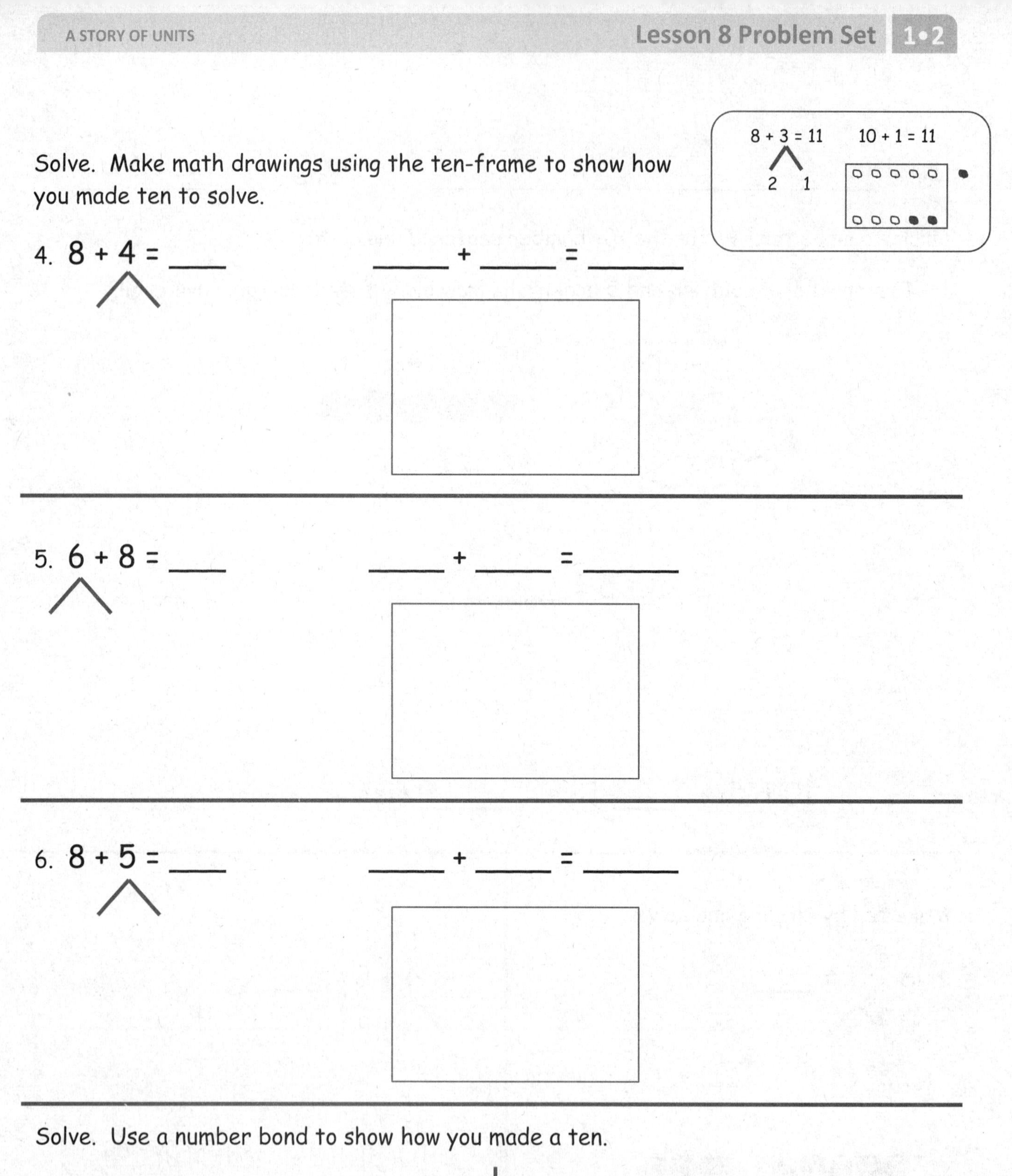

Solve. Make math drawings using the ten-frame to show how you made ten to solve.

4. 8 + 4 = ____ ____ + ____ = ____

5. 6 + 8 = ____ ____ + ____ = ____

6. 8 + 5 = ____ ____ + ____ = ____

Solve. Use a number bond to show how you made a ten.

7. 5 + 8 = ____

8. ____ = 8 + 7

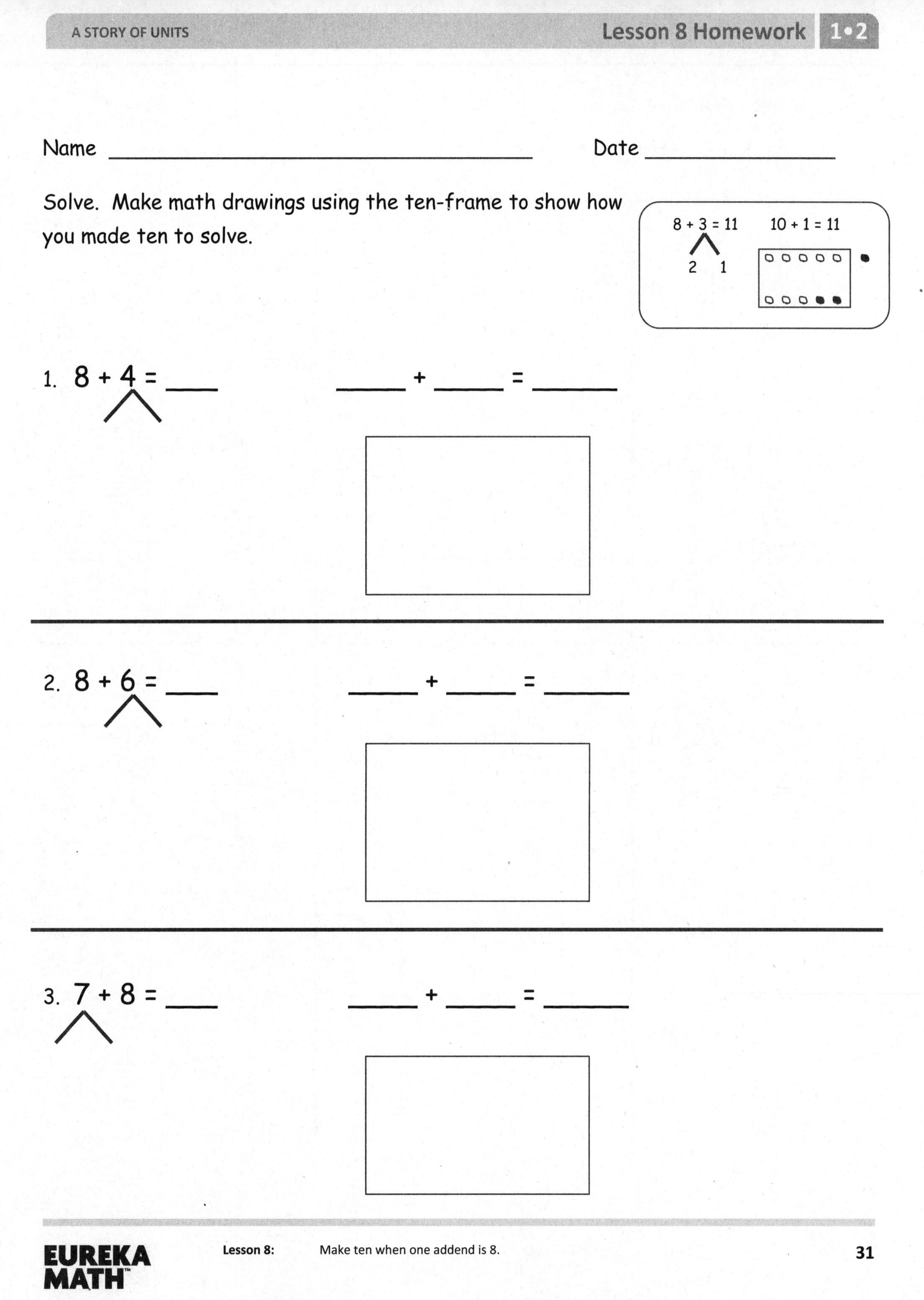

Name ______________________ Date ____________

Solve. Make math drawings using the ten-frame to show how you made ten to solve.

$8 + 3 = 11$ (3 split into 2 and 1) $10 + 1 = 11$

1. $8 + 4 =$ ___ ___ + ___ = ___

2. $8 + 6 =$ ___ ___ + ___ = ___

3. $7 + 8 =$ ___ ___ + ___ = ___

4. Make math drawings using ten-frames to solve. Circle the true number sentences.

Write an X to show number sentences that are not true.

a. $8 + 4 = 10 + 2$

b. $10 + 6 = 8 + 8$

c. $7 + 8 = 10 + 6$

d. $5 + 10 = 5 + 8$

e. $2 + 10 = 8 + 3$

f. $8 + 9 = 10 + 7$

Name ______________________ Date ____________

Make ten to solve. Use a number bond to show how you took 2 out to make ten.

1. Ben has 8 green grapes and 3 purple grapes. How many grapes does he have?

8 + 3 = ____   10 + ____ = ____

Ben has ___ grapes.

---

2. 8 + 4 = ____   10 + ____ = ____

Use number bonds to show your thinking. Write the 10+ fact.

3. 8 + 5 = ____   ____ + ____ = ____

4. 8 + 7 = ____   ____ + ____ = ____

5. 4 + 8 = ____   ____ + ____ = ____

6. 7 + 8 = ____   ____ + ____ = ____

7. 8 + ____ = 17   ____ + ____ = ____

Complete the addition sentences and number bonds.

8. a. 10 + 1 = ____ (number bond: 11) b. 8 + 3 = ____ (number bond: 11)

9. a. 10 + 5 = ____ (number bond: 15) b. 8 + 7 = ____ (number bond: 15)

10. a. 10 + 6 = ____ b. 8 + 8 = ____

11. a. 2 + 10 = ____ b. 4 + 8 = ____

12. a. 4 + 10 = ____ b. 6 + 8 = ____

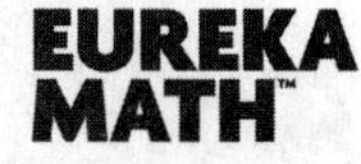

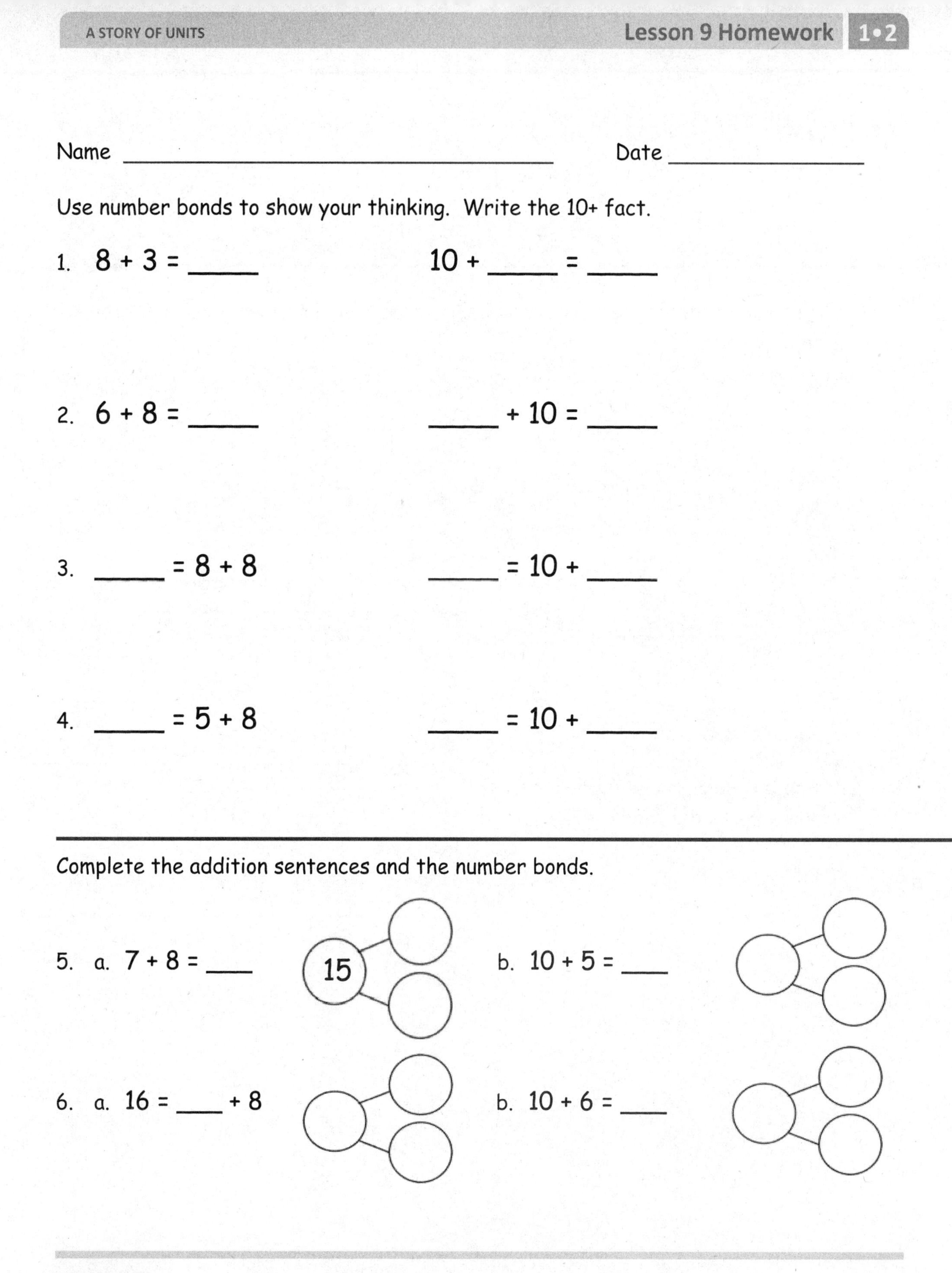

Name ______________________________ Date ______________

Use number bonds to show your thinking. Write the 10+ fact.

1. 8 + 3 = _____ 10 + _____ = _____

2. 6 + 8 = _____ _____ + 10 = _____

3. _____ = 8 + 8 _____ = 10 + _____

4. _____ = 5 + 8 _____ = 10 + _____

Complete the addition sentences and the number bonds.

5. a. 7 + 8 = ____ b. 10 + 5 = ____

6. a. 16 = ____ + 8 b. 10 + 6 = ____

7. a. ____ = 9 + 8

b. 10 + 7 = ____

Draw a line to the matching number sentence. You may use a number bond or 5-group drawing to help you.

8. 11 = 8 + 3

8 + 6 = 14

9. Lisa had 5 red rocks and 8 white rocks.
How many rocks did she have?

10 + 1 = 11

13 = 10 + 3

10.

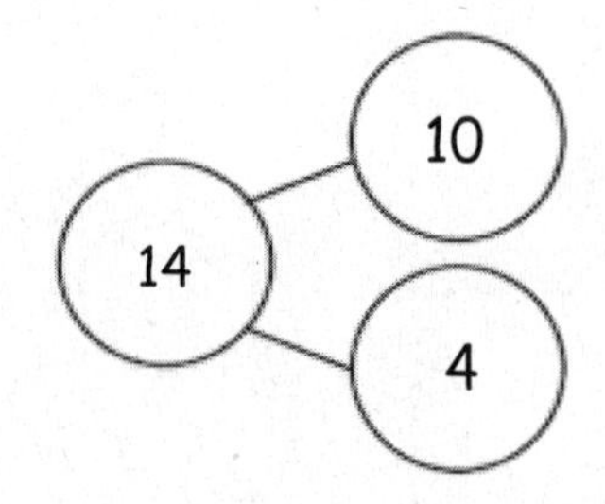

Name ______________________________ Date ______________

Solve. Use number bonds or 5-group drawings if needed. Write the equal ten-plus number sentence.

1. 4 + 9 = ___

10 + ___ = ___

2. 6 + 8 = ___

10 + ___ = ___

3. 7 + 4 = ___

10 + ___ = ___

4. Match the equal expressions.

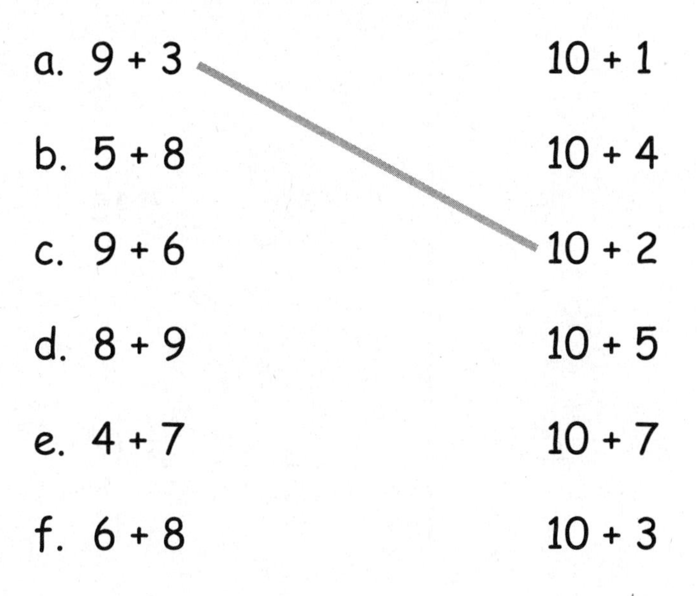

a. 9 + 3	10 + 1
b. 5 + 8	10 + 4
c. 9 + 6	10 + 2
d. 8 + 9	10 + 5
e. 4 + 7	10 + 7
f. 6 + 8	10 + 3

Complete the addition sentences to make them true.

	a.	b.	c.
5.	$9 + 2 =$ ___	$8 + 4 =$ ___	$7 + 5 =$ ___
6.	$9 + 5 =$ ___	$8 + 3 =$ ___	$7 + 6 =$ ___
7.	$6 + 9 =$ ___	$6 + 8 =$ ___	$4 + 7 =$ ___
8.	$7 + 9 =$ ___	$5 + 8 =$ ___	$7 + 7 =$ ___
9.	$9 +$ ___ $= 17$	$8 +$ ___ $= 16$	$7 +$ ___ $= 16$
10.	___ $+ 9 = 15$	___ $+ 8 = 15$	___ $+ 7 = 17$

Name ______________________________ Date ______________

Solve. Match the number sentence to the ten-plus number bond that helped you solve the problem. Write the ten-plus number sentence.

1. 8 + 6 = ____

____ + ____ = ____

2. 7 + 5 = ____

____ + ____ = ____

3. 5 + 8 = ____

____ + ____ = ____

4. 4 + 7 = ____

____ + ____ = ____

5. 6 + 9 = ____

____ + ____ = ____

Complete the number sentences so they equal the given number bond.

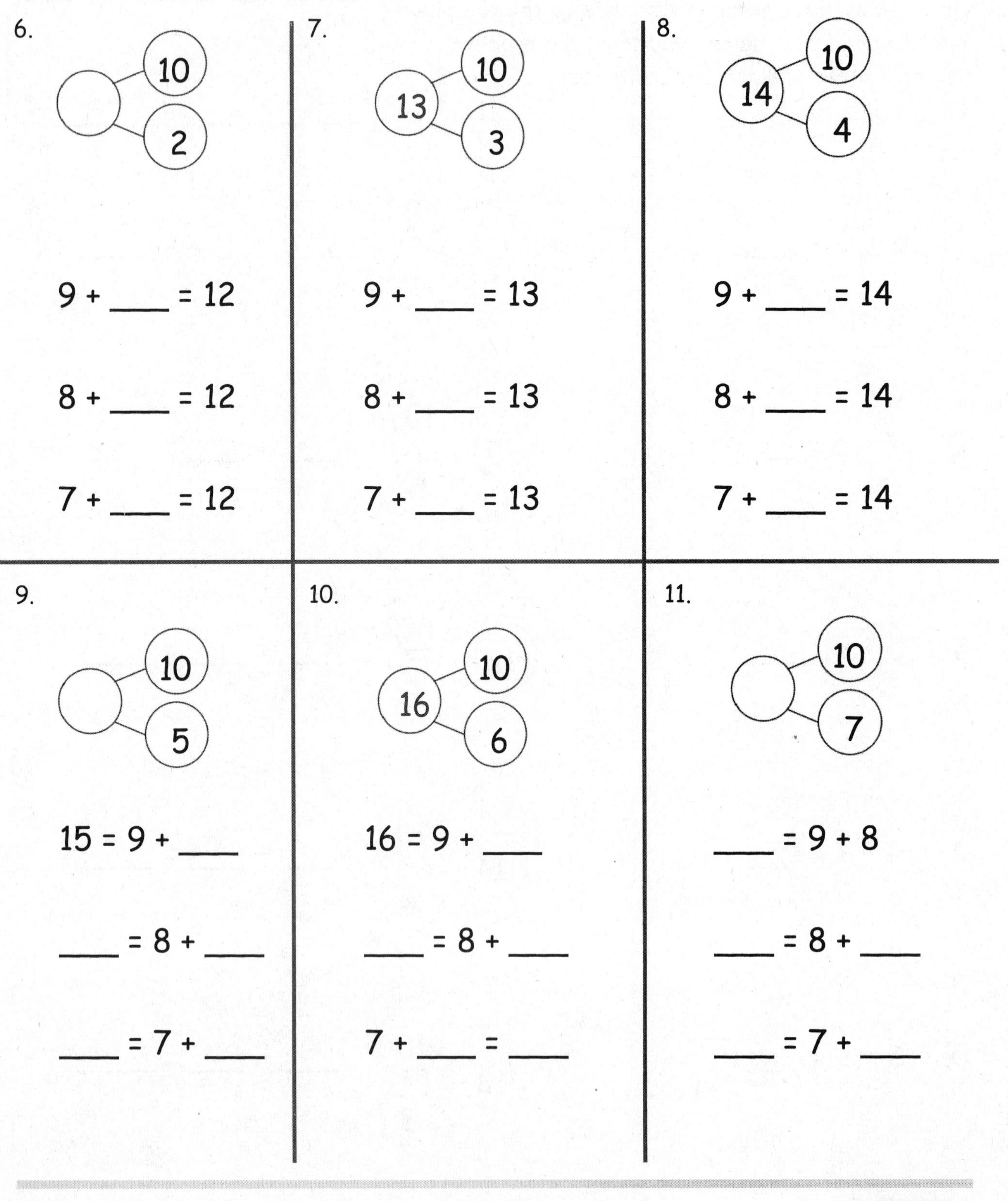

Name ______________________________ Date ______________

Jeremy had 7 big rocks and 8 little rocks in his pocket.

How many rocks does Jeremy have?

1. Circle all student work that correctly matches the story.

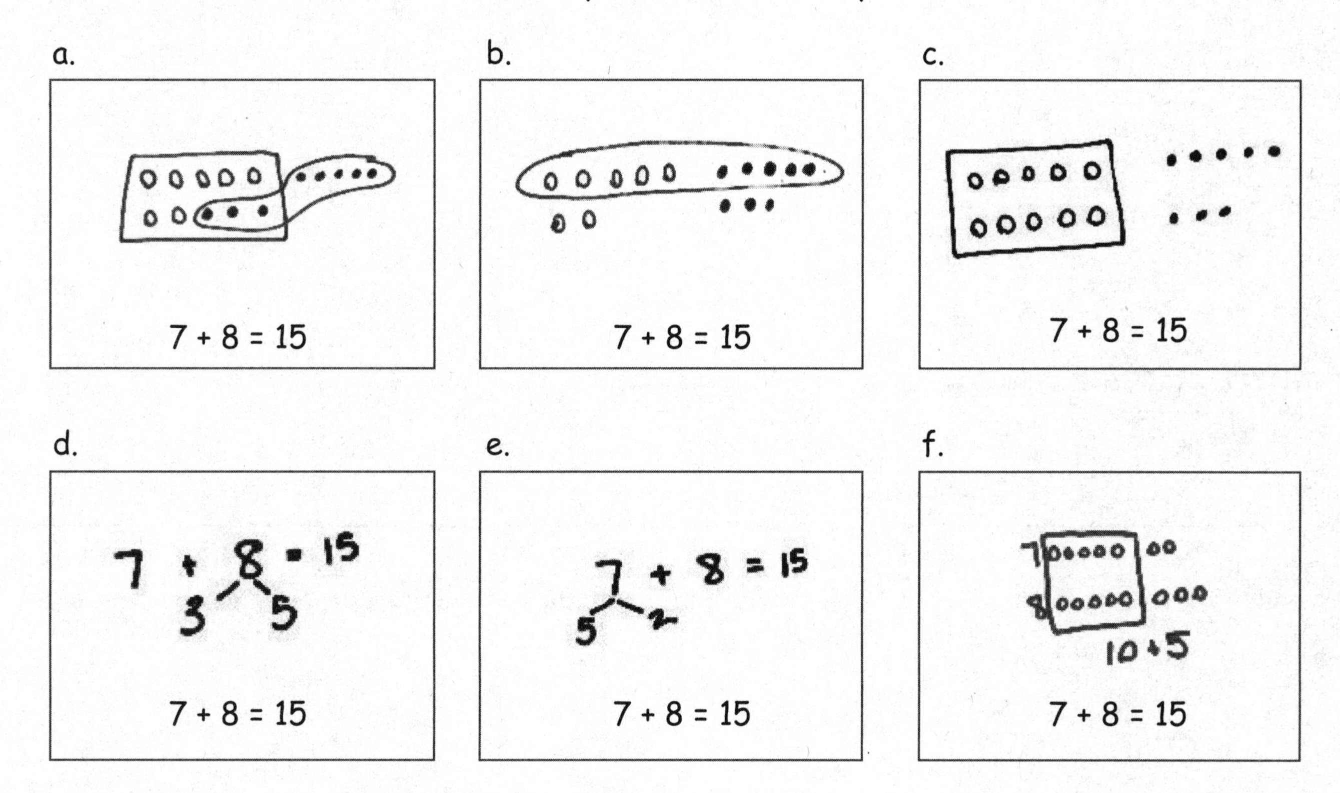

2. Fix the work that was incorrect by making a new drawing in the space below with the matching number sentence.

Solve on your own. Show your thinking by drawing or writing. Write a statement to answer the question.

3. There are 4 vanilla cupcakes and 8 chocolate cupcakes for the party. How many cupcakes were made for the party?

---

4. There are 5 girls and 7 boys on the playground. How many students are on the playground?

When you are done, share your solutions with a partner. How did your partner solve each problem? Be ready to share how your partner solved the problems.

Name ______________________________ Date ______________

Look at the student work. Correct the work. If the answer is incorrect, show a correct solution in the space below the student work.

1. Todd has 9 red cars and 7 blue cars. How many cars does he have altogether?

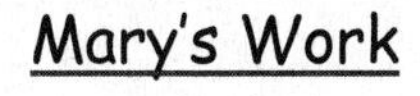

Mary's Work Joe's Work Len's Work

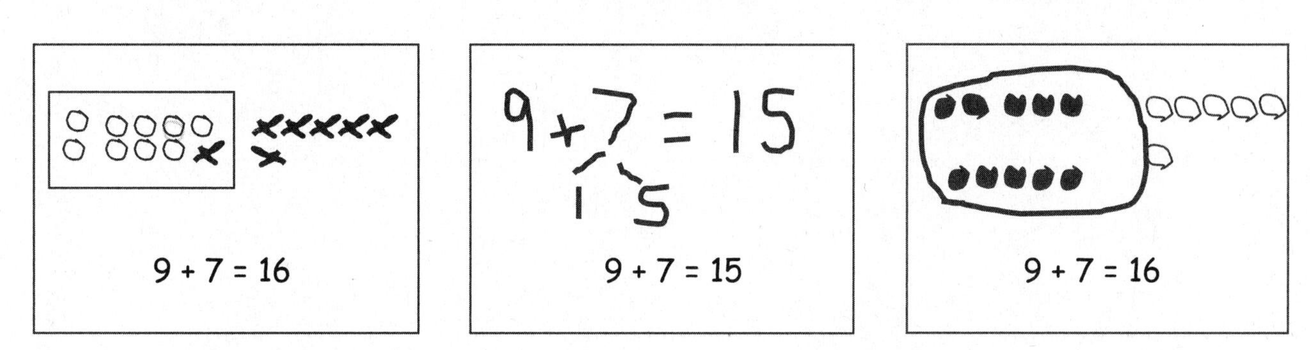

2. Jill has 8 beta fish and 5 goldfish. How many fish does she have in total?

Frank's Work Lori's Work Mike's Work

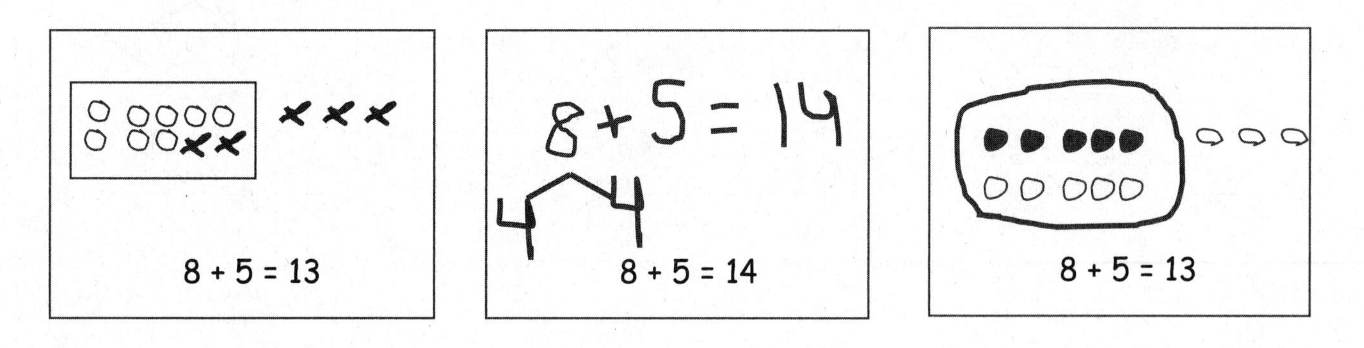

3. Dad baked 7 chocolate and 6 vanilla cupcakes. How many cupcakes did he bake in all?

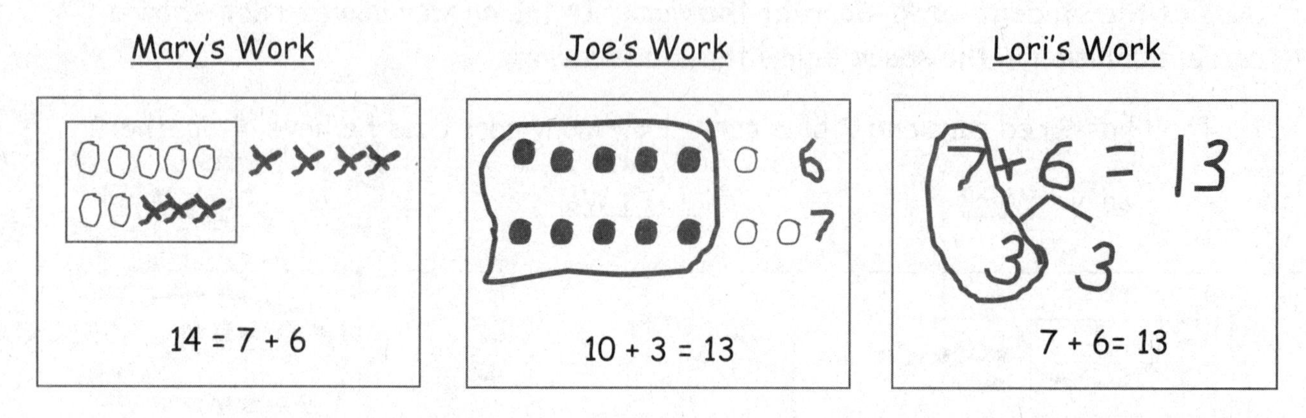

4. Mom caught 9 fireflies, and Sue caught 8 fireflies. How many fireflies did they catch altogether?

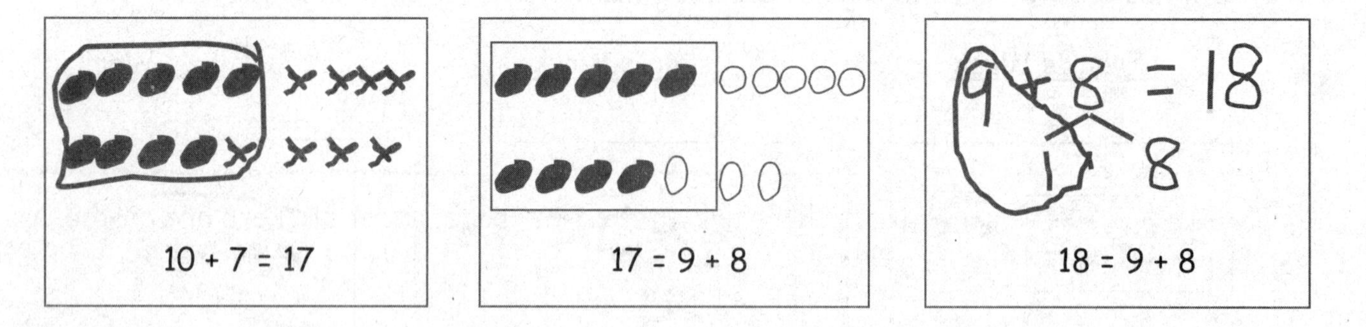

Name ______________________________ Date ________________

Make a simple math drawing. Cross out from the 10 ones or the other part in order to show what happens in the stories.

1. Bill has 16 grapes. 10 are on one vine, and 6 are on the ground. Bill eats 9 grapes from the vine. How many grapes does Bill have left?

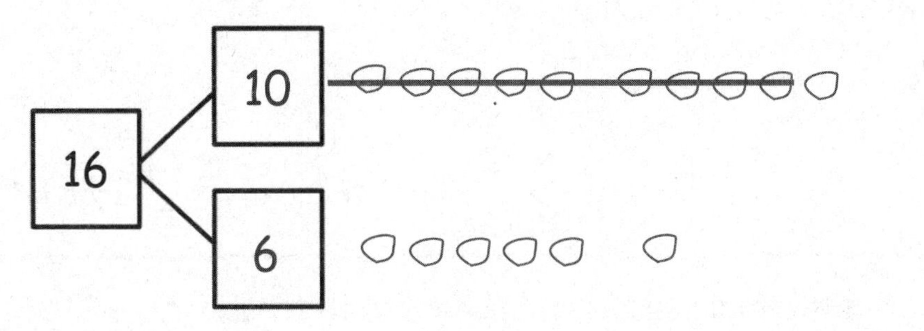

Bill has _____ grapes now.

2. 12 frogs are in the pond. 10 are on a lily pad, and 2 are in the water. 9 frogs hop off the lily pad and out of the pond. How many frogs are in the pond?

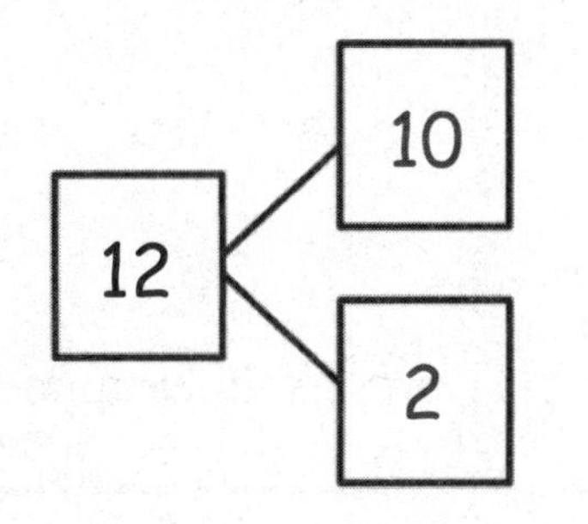

There are _____ frogs still in the pond.

3. Kim has 14 stickers. 10 stickers are on the first page, and 4 stickers are on the second page. Kim loses 9 stickers from the first page. How many stickers are still in her book?

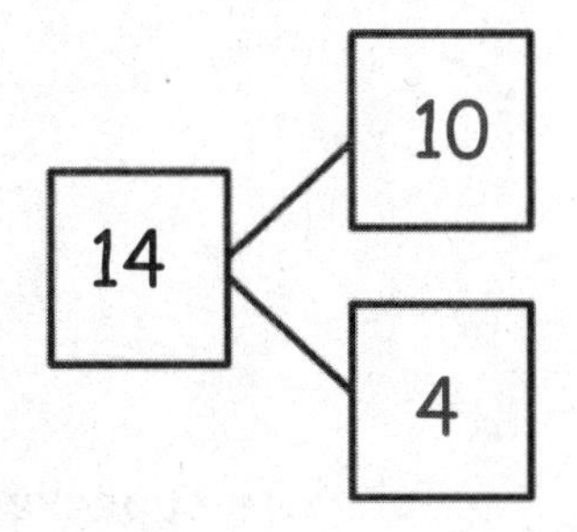

Kim has _____ stickers in her book.

4. 10 eggs are in a carton, and 5 eggs are in a bowl. Joe's father cooks 9 eggs from the carton. How many eggs are left?

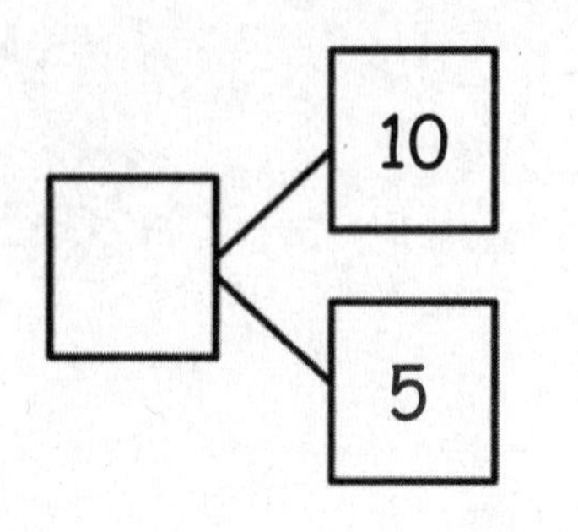

There are ___ eggs left.

---

5. Jana had 10 wrapped gifts on the table and 7 wrapped gifts on the floor. She unwrapped 9 gifts from the table. How many gifts are still wrapped?

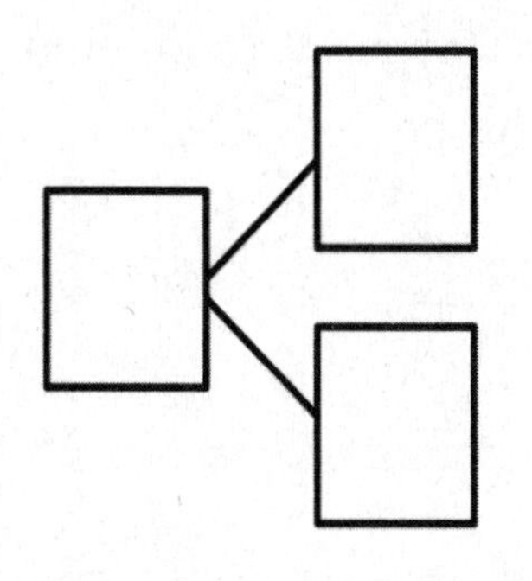

Jana has ___ gifts still wrapped.

---

6. There are 10 cupcakes on a tray and 8 on the table. On the tray, there are 9 vanilla cupcakes. The rest of the cupcakes are chocolate. How many cupcakes are chocolate?

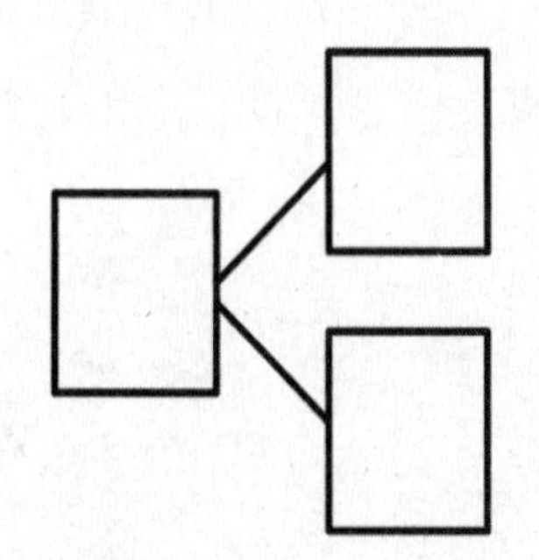

There are ___ chocolate cupcakes.

Name ______________________________ Date ______________

Make a simple math drawing. Cross out from the 10 ones to show what happens in the stories.

I had 16 grapes.
10 of them were red,
and 6 were green.
I ate 9 red grapes.
How many grapes do
I have now?

16 → 10, 6

Now I have 7 grapes.

1. There were 15 squirrels by a tree. 10 of them were eating nuts. 5 squirrels were playing. A loud noise scared away 9 of the squirrels eating nuts. How many squirrels were left by the tree?

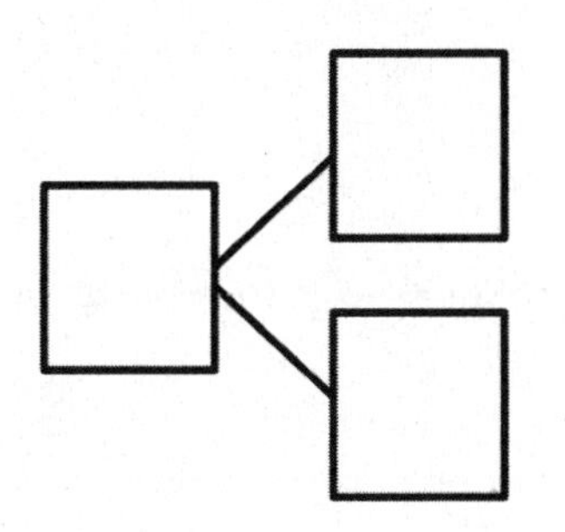

There were ___ squirrels left by the tree.

2. There are 17 ladybugs on the plant. 10 of them are on a leaf, and 7 of them are on the stem. 9 of the ladybugs on the leaf crawled away. How many ladybugs are still on the plant?

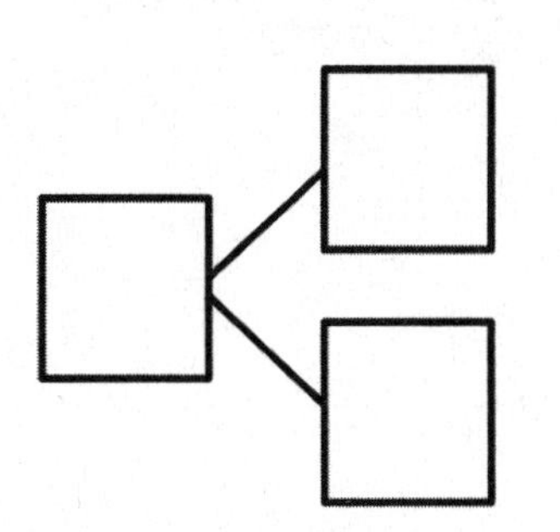

There are ___ ladybugs on the plant.

3. Use the number bond to fill in the math story. Make a simple math drawing. Cross out from 10 ones or some ones to show what happens in the stories.

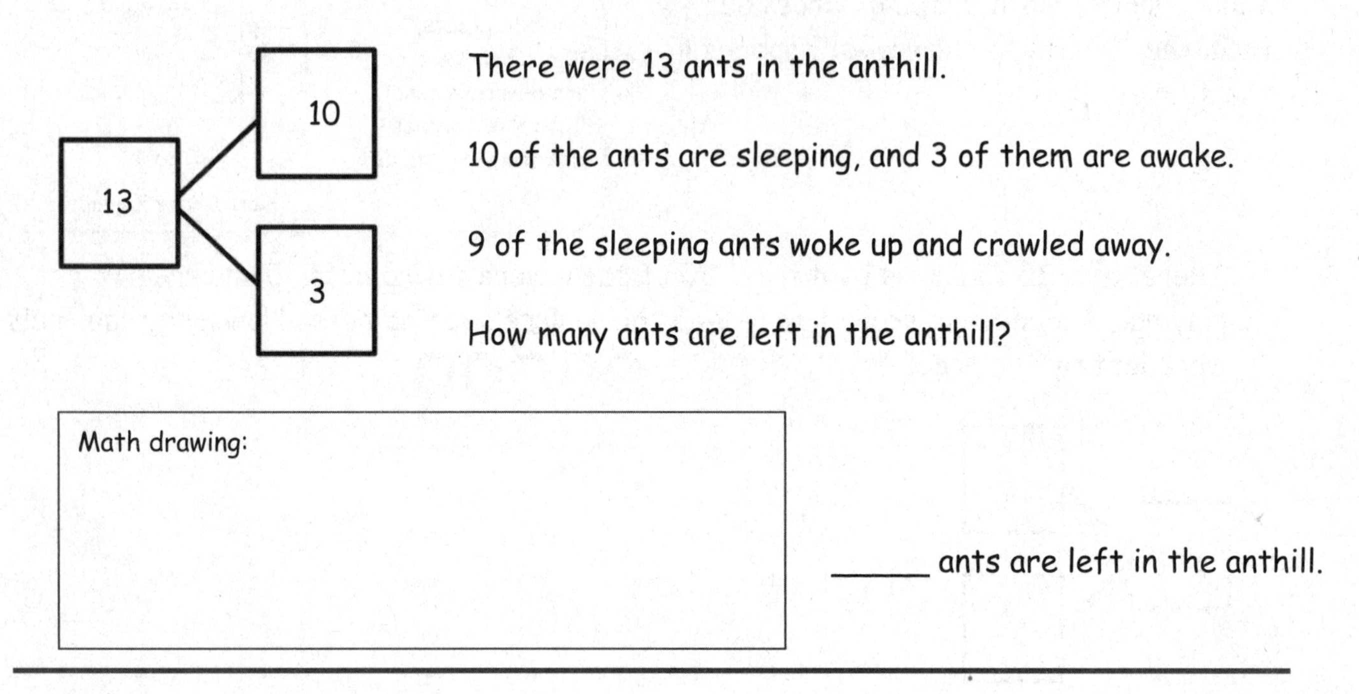

There were 13 ants in the anthill.

10 of the ants are sleeping, and 3 of them are awake.

9 of the sleeping ants woke up and crawled away.

How many ants are left in the anthill?

Math drawing:

_____ ants are left in the anthill.

---

4. Use the number bond below to come up with your own math story. Include a simple math drawing. Cross out from 10 ones to show what happens.

14

10

4

Math drawing:

Number sentences:

Statement:

00000 00000

5-group row insert

This page intentionally left blank

Name ______________________ Date ____________

Solve. Use 5-group rows, and cross out to show your work.

1. Mike has 10 cookies on a plate and 3 cookies in a box. He eats 9 cookies from the plate. How many cookies are left?

13
10 3

○○○○○ ○○○○○ ●●●

Mike has ___ cookies left.

---

2. Fran has 10 crayons in a box and 5 crayons on the desk. Fran lends Bob 9 crayons from the box. How many crayons does Fran have to use?

15

Fran has ___ crayons to use.

---

3. 10 ducks are in the pond, and 7 ducks are on the land. 9 of the ducks in the pond are babies, and all the rest of the ducks are adults. How many adult ducks are there?

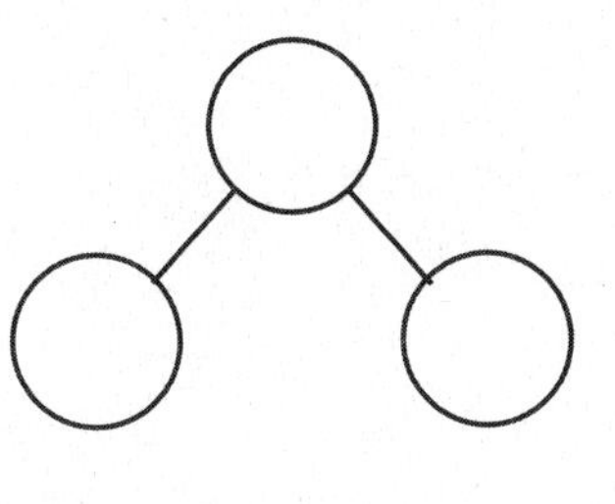

There are ___ adult ducks.

With a partner, create your own stories to match, and solve the number sentences. Make a number bond to show the whole as 10 and some ones. Draw 5-group rows to match your story. Write the complete number sentence on the line.

4. 16 - 9 = ☐

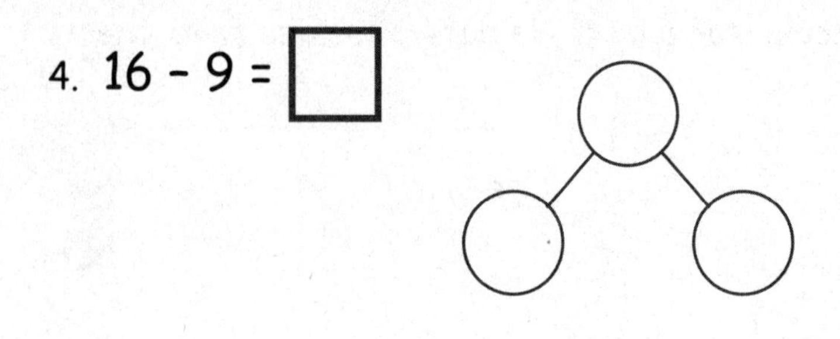

____

5. 12 - 9 = ☐

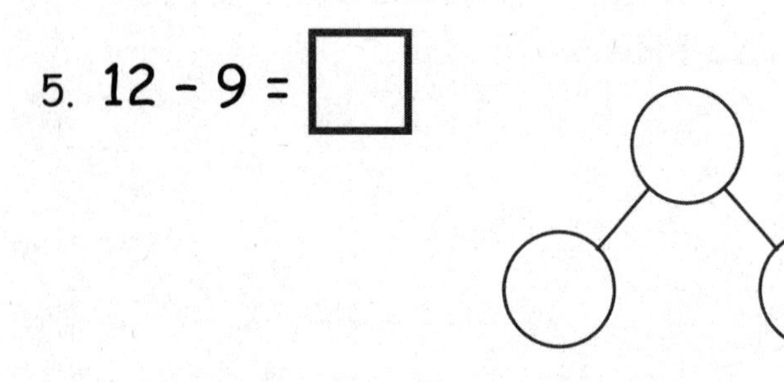

____

6. 19 - 9 = ☐

____

Name ______________________________ Date ______________

Solve. Use 5-group rows, and cross out to show your work. Write number sentences.

13

10 3

13 - 9 = 4

1. In a park, 10 dogs are running on the grass, and 1 dog is sleeping under the tree. 9 of the running dogs leave the park. How many dogs are left in the park?

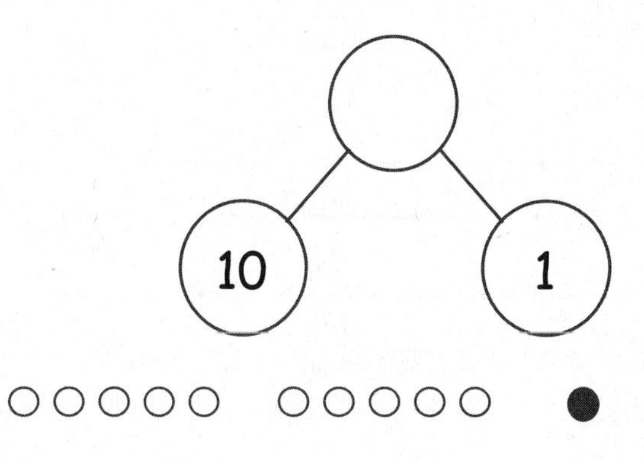

There are ___ dogs left in the park.

---

2. Alejandro had 9 rocks in his yard and 10 rocks in his room. 9 of the rocks in his room are gray rocks, and the rest of the rocks are white. How many white rocks does Alejandro have?

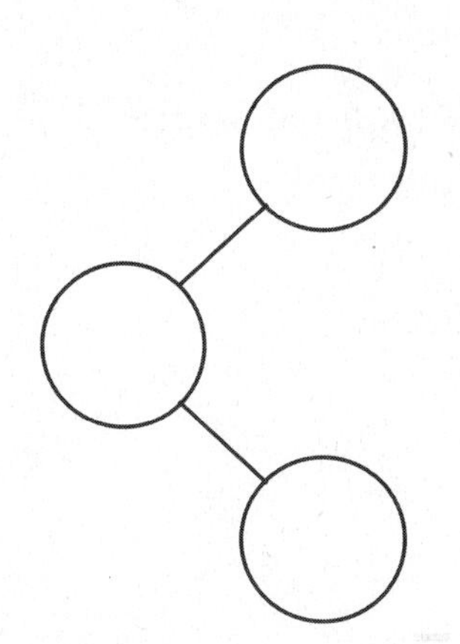

Alejandro has ___ white rocks.

G1-M2-SE-B2-1.3.1-12.2015

3. Sophia has 8 toy cars in the kitchen and 10 toy cars in her bedroom. 9 of the toy cars in the bedroom are blue. The rest of her cars are red. How many red cars does Sophia have?

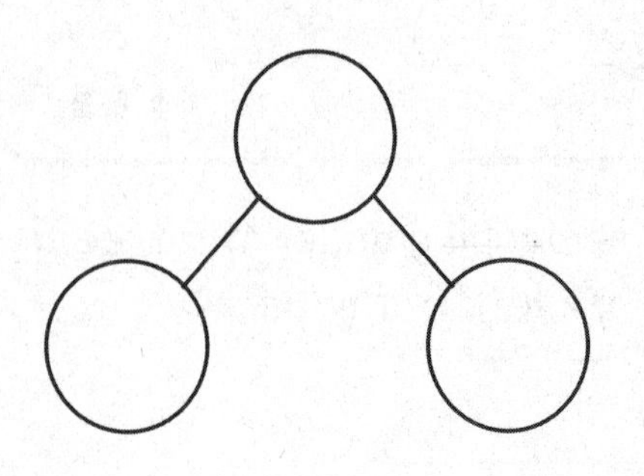

Sophia has ___ red cars.

---

4. Complete the number bond, and fill in the math story. Use 5-group rows, and cross out to show your work. Write number sentences.

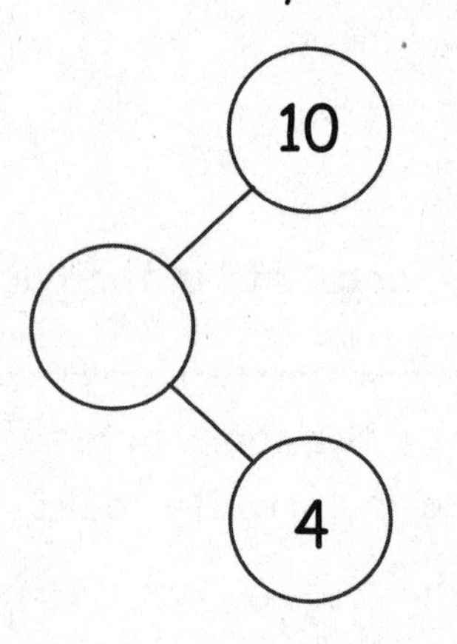

There were ____ birds splashing in a puddle and ____ birds walking on the dry grass. 9 of the splashing birds flew away. How many birds are left?

There are ___ birds left.

Name ______________________________ Date ______________

1. Match the pictures with the number sentences.

a. 11 - 9 = 2

b. 14 - 9 = 5

c. 16 - 9 = 7

d. 18 - 9 = 9

e. 17 - 9 = 8

Circle 10 and subtract.

2. 12 - 9 = _____

3. 14 - 9 = _____

4. 15 - 9 = ____

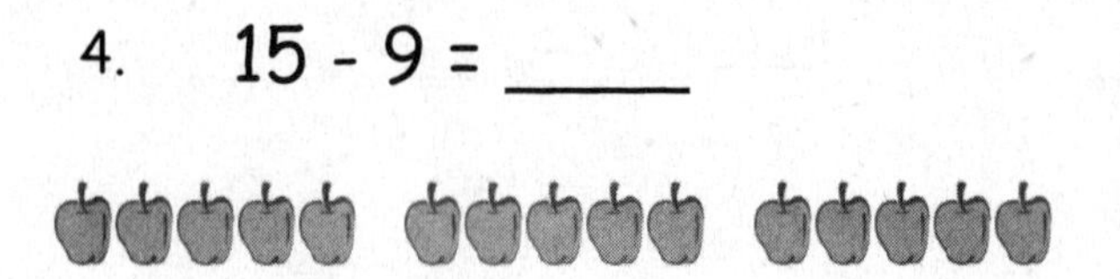

5. 13 - 9 = ____

6. 16 - 9 = ____

7. 17 - 9 = ____

Draw and circle 10. Then subtract.

8. 12 - 9 = ____

9. 13 - 9 = ____

10. 14 - 9 = ____

11. 15 - 9 = ____

Name GABRIEL Date 2020

Circle 10 and subtract. Make a number bond.

1. 15 - 9 = 6

Draw and circle 10. Subtract and make a number bond.

2. 14 - 9 = 5

3. 12 - 9 = 3

4. 13 - 9 = 4

5. 16 - 9 = 7

6. Complete the number bond, and write the number sentence that helped you.

7. Make the number bond that would come next, and write a number sentence that matches.

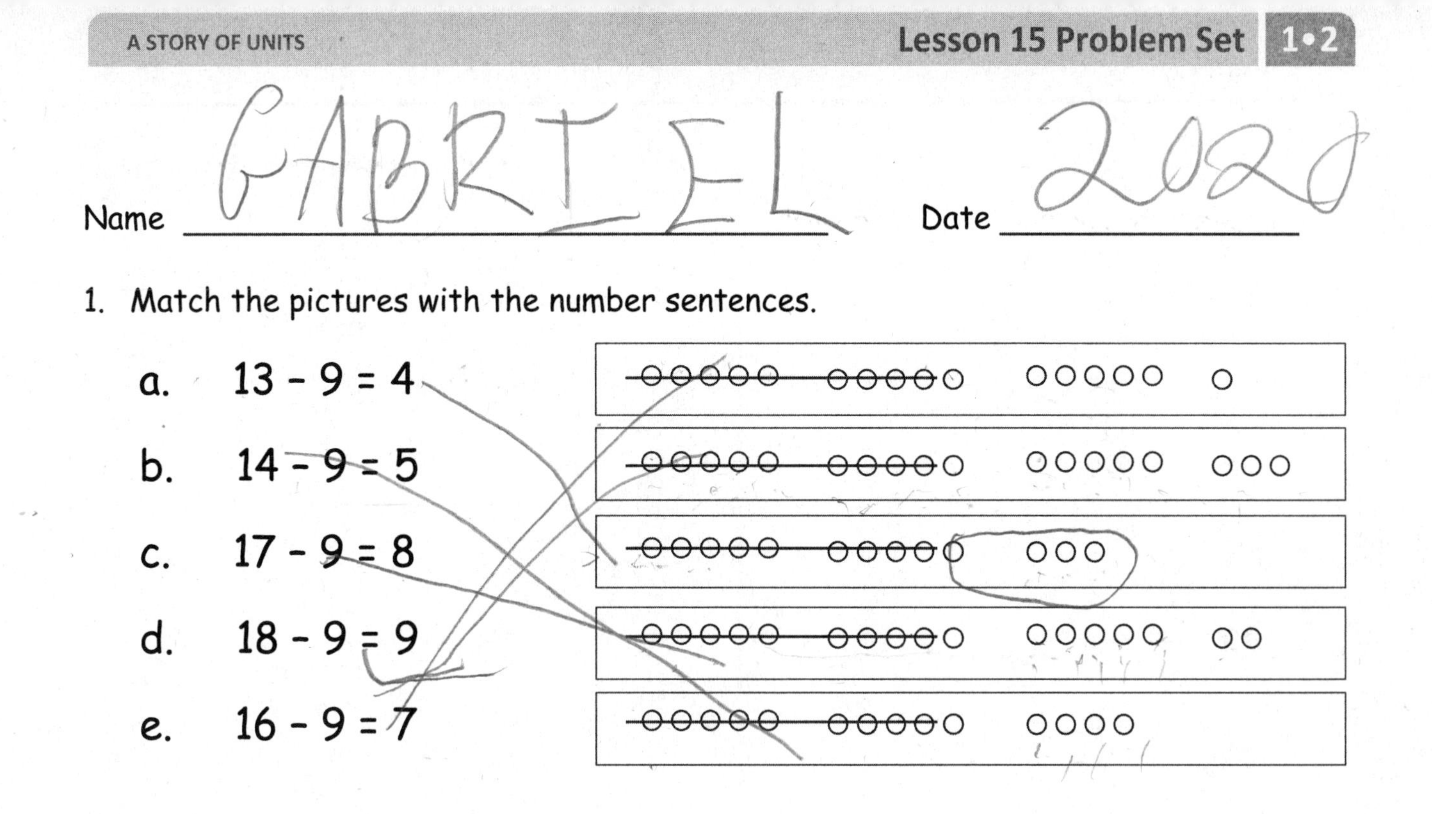

Name GABRIEL Date 2020

1. Match the pictures with the number sentences.

   a. 13 - 9 = 4

   b. 14 - 9 = 5

   c. 17 - 9 = 8

   d. 18 - 9 = 9

   e. 16 - 9 = 7

Draw 5-group rows. Visualize and then cross out to solve. Complete the number sentences.

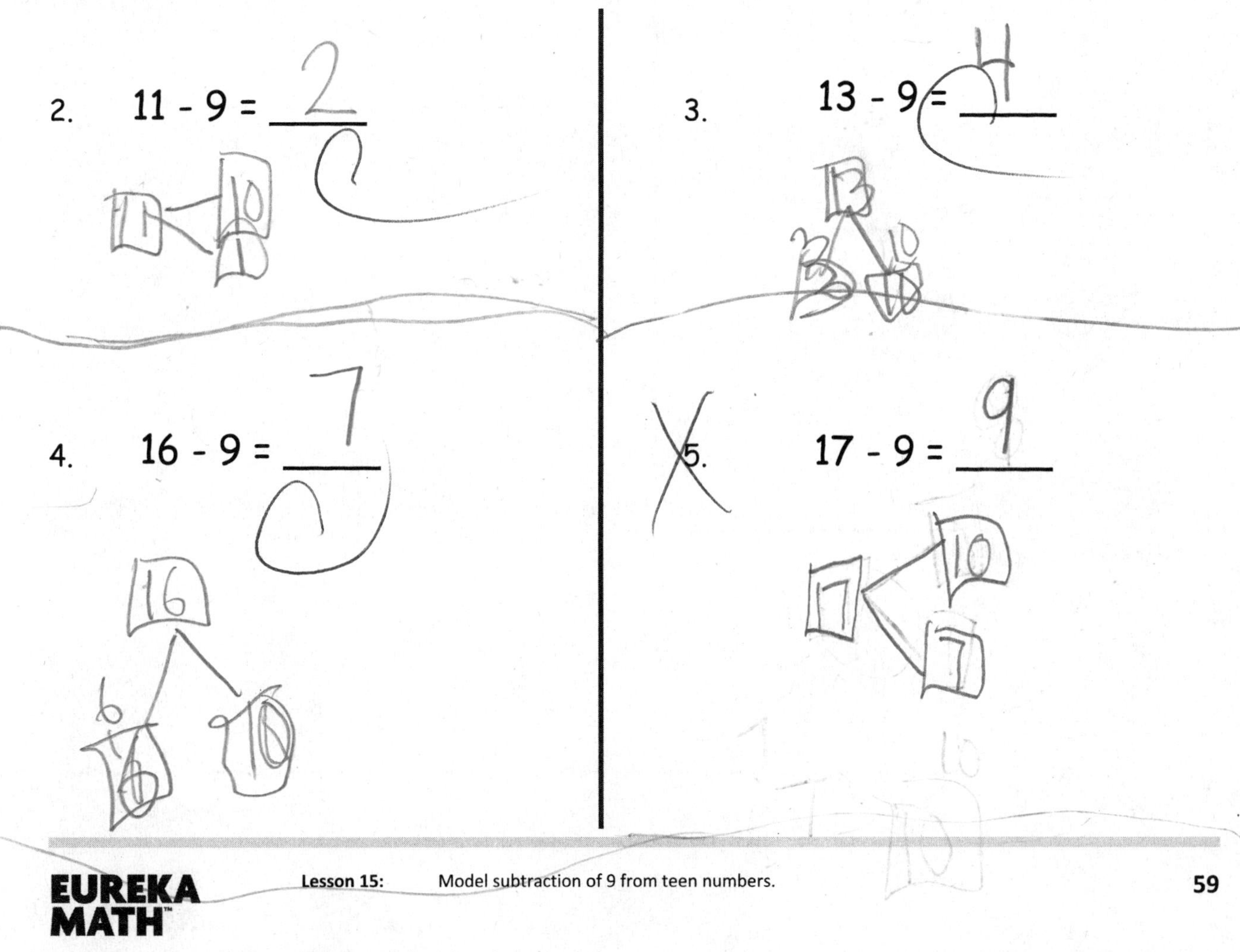

2. 11 - 9 = 2

3. 13 - 9 = 4

4. 16 - 9 = 7

5. 17 - 9 = 9

GABRIEL

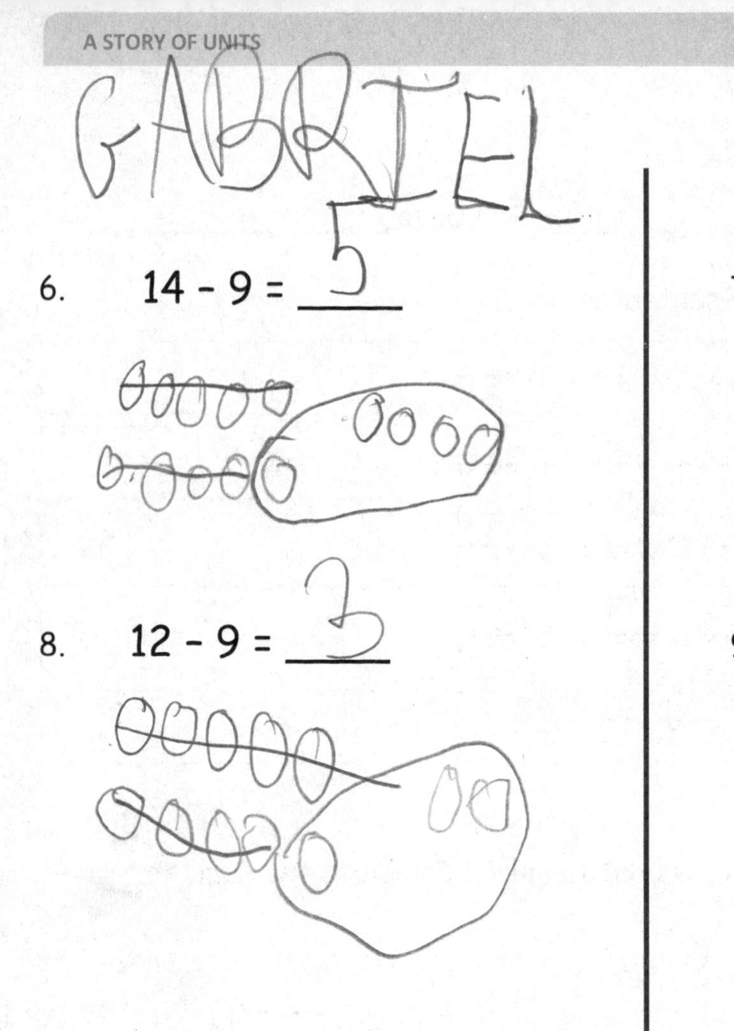

6. $14 - 9 =$ 5

7. $13 - 9 =$ 4

8. $12 - 9 =$ 3

9. $15 - 9 =$ 6

10. Show making 10 and taking from 10 to complete the two number sentences.

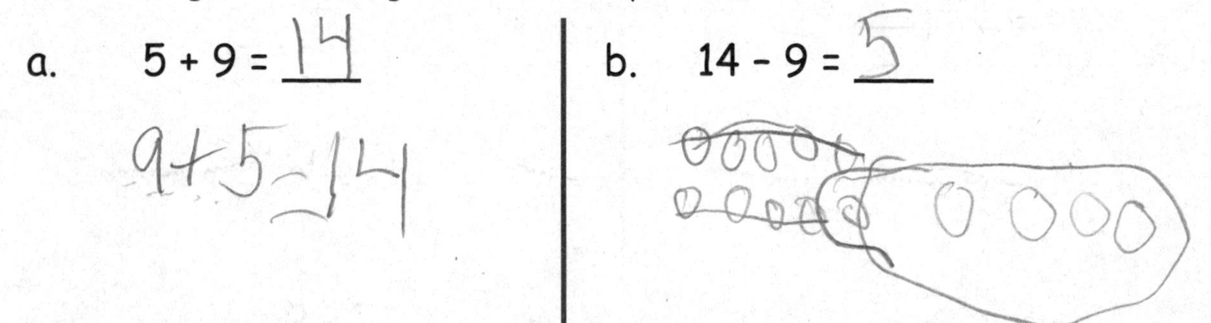

a. $5 + 9 =$ 14

9+5=14

b. $14 - 9 =$ 5

11. Make a number bond for Problem 10. Write two additional number sentences that use this number bond.

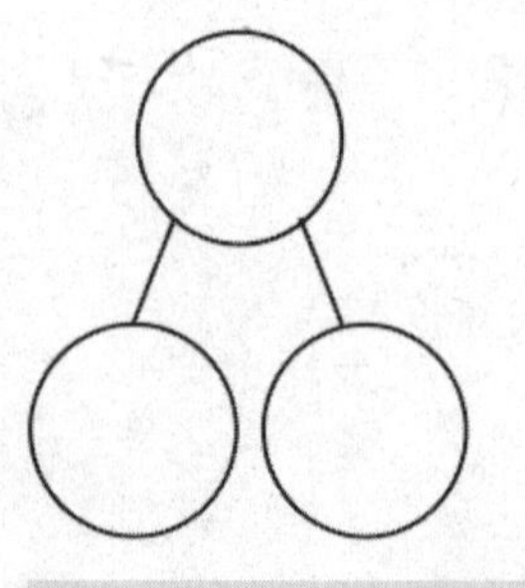

_______________ _______________

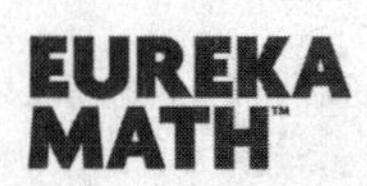

Name ________________ Date ____________

Write the number sentence for each 5-group row drawing.

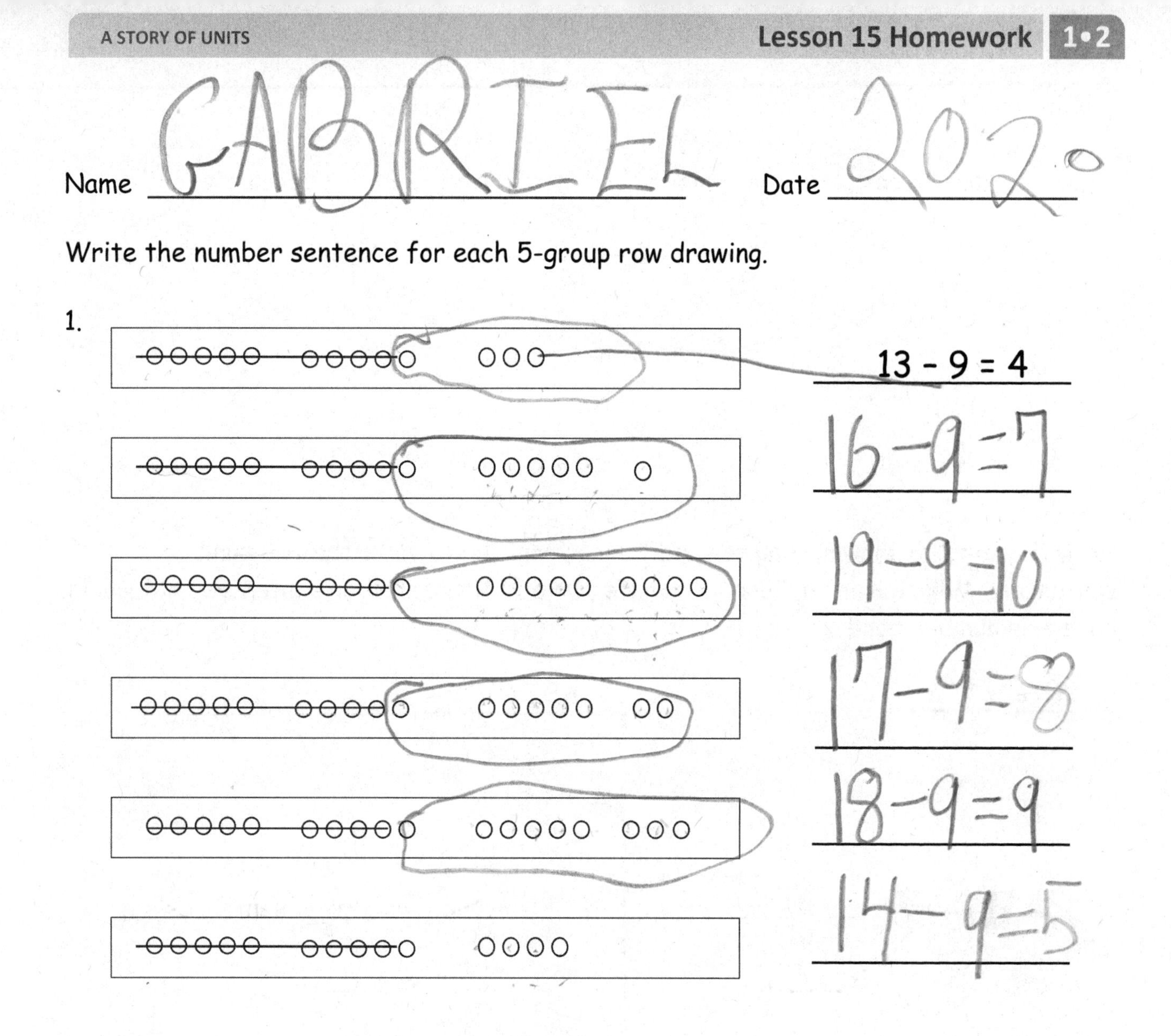

1. 13 - 9 = 4

Draw 5-groups to complete the number bond, and write the 9- number sentence.

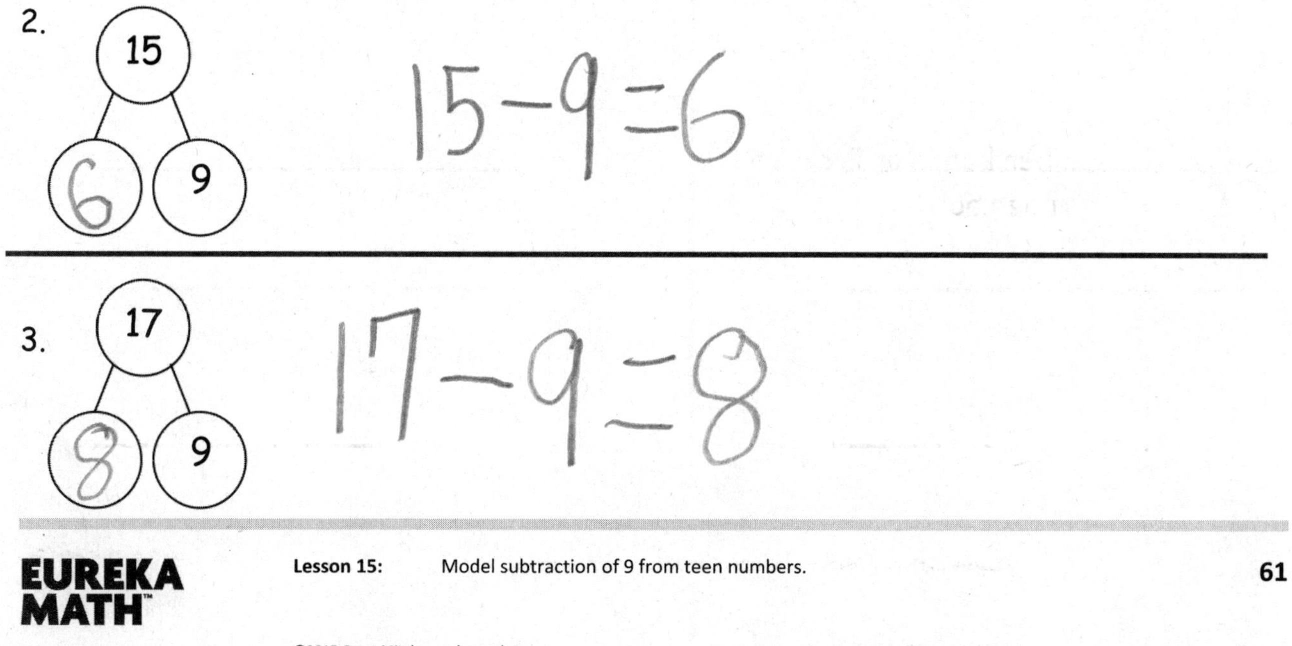

Draw 5-groups to complete the number bond, and write the 9- number sentence.

4.

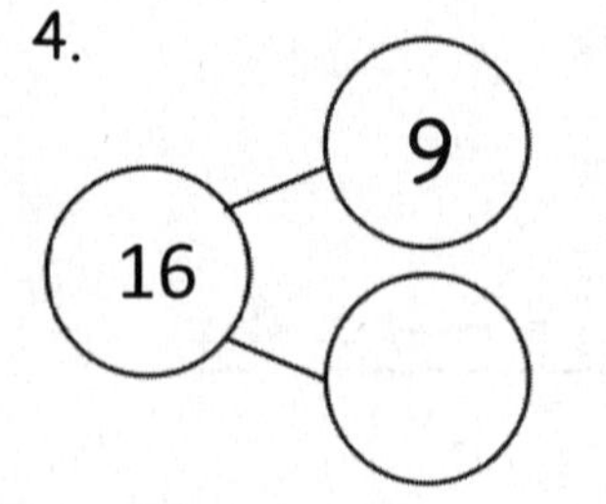

Draw 5-groups to show making ten and taking from ten to solve the two number sentences. Make a number bond, and write two additional number sentences that would have this number bond.

5. 8 + 9 = ____

6. 17 − 9 = ____

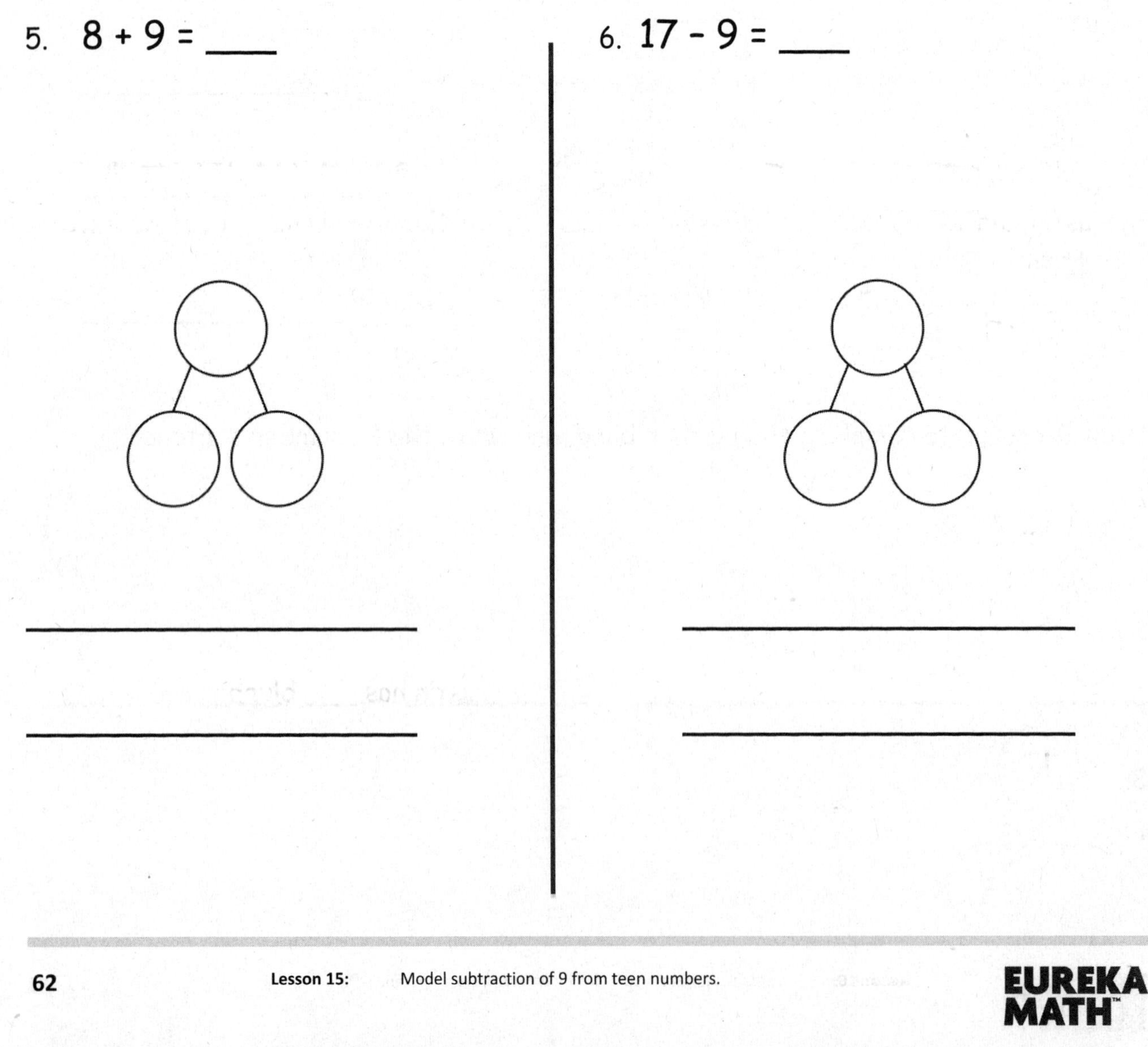

Name ______________________ Date __________

Solve the problem by counting on (a) and using a number bond to take from ten (b).

1. Lucy had 12 balloons at her birthday party. She gave 9 balloons to her friends. How many balloons did she have left?

   a. 12 - 9 = ____

   b. 12 - 9 = ____

Lucy had ___ balloons left.

---

2. Justin had 15 blueberries on his plate. He ate 9 of them. How many does he have left to eat?

   a. 15 - 9 = ____

   b. 15 - 9 = ____

Justin has ___ blueberries left to eat.

Complete the subtraction sentences by using the take from ten strategy and counting on. Tell which strategy you would prefer to use for Problems 3 and 4.

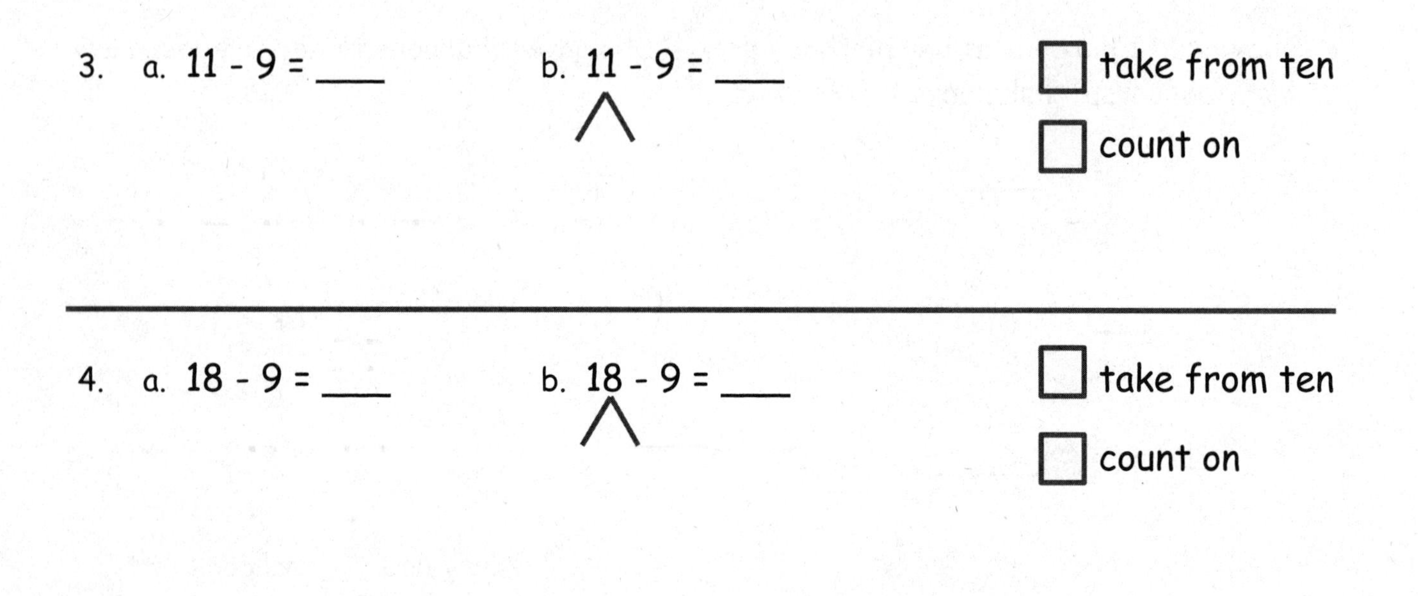

3. a. 11 - 9 = ___ b. 11 - 9 = ___

☐ take from ten

☐ count on

4. a. 18 - 9 = ___ b. 18 - 9 = ___

☐ take from ten

☐ count on

5. Think about how to solve the following subtraction problems:

16 - 9	12 - 9	18 - 9
11 - 9	15 - 9	14 - 9
13 - 9	19 - 9	17 - 9

Choose which problems you think are easier to count on from 9 and which are easier to use the take from ten strategy. Write the problems in the boxes below.

Problems to use the *count on* strategy with:	**Problems to use the *take from ten* strategy with:**

Were there any problems that were just as easy using either method? Did you use a different method for any problems?

Name ______________________________ Date ______________

Complete the subtraction sentences by using either the count on or take from ten strategy. Tell which strategy you used.

1. 17 - 9 = ___

☐ take from ten

☐ count on

2. 12 - 9 = ___

☐ take from ten

☐ count on

3. 16 - 9 = ___

☐ take from ten

☐ count on

4. 11 - 9 = ___

☐ take from ten

☐ count on

5. Nicholas collected 14 leaves. He pasted 9 into his notebook. How many of his leaves were not pasted into his notebook? Choose the count on or take from ten strategy to solve.

I chose this strategy:

☐ take from ten

☐ count on

6. Sheila had 17 oranges. She gave 9 oranges to her friends. How many oranges does Sheila have left? Choose the count on or take from ten strategy to solve.

I chose this strategy:

☐ take from ten

☐ count on

7. Paul has 12 marbles. Lisa has 18 marbles. They each rolled 9 marbles down a hill. How many marbles did each student have left? Tell which strategy you chose for each student.

Paul has ____ marbles left. Lisa has ____ marbles left.

8. Just as you did today in class, think about how to solve the following problems, and talk to your parent or caregiver about your ideas.

15 - 9	13 - 9	17 - 9
18 - 9	19 - 9	12 - 9
11 - 9	14 - 9	16 - 9

Circle the problems you think are easier to solve by counting on from 9. Put a rectangle around those that are easier to solve using the take from ten strategy. Remember, some might be just as easy using either method.

Name ______________________________ Date ______________

1. Match the pictures with the number sentences.

a. 12 - 8 = 4

b. 17 - 8 = 9

c. 16 - 8 = 8

d. 18 - 8 = 10

e. 14 - 8 = 6

Circle 10 and subtract.

2. 13 - 8 = _____

3. 11 - 8 = _____

4. 15 - 8 = ____

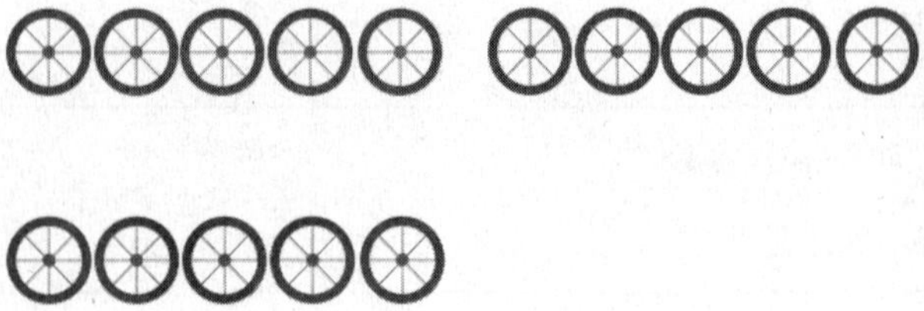

5. 19 - 8 = ____

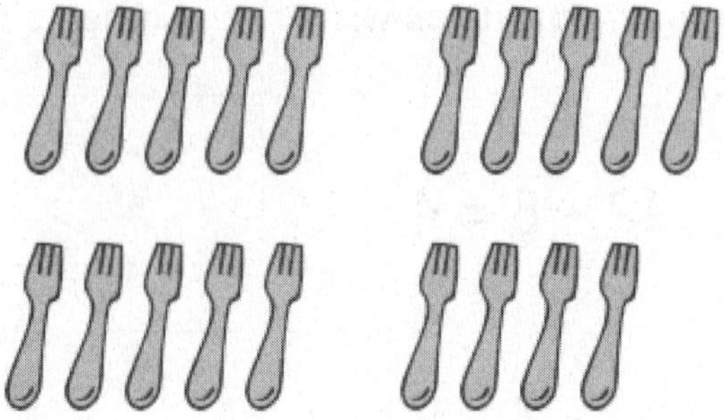

6. 16 - 8 = ____

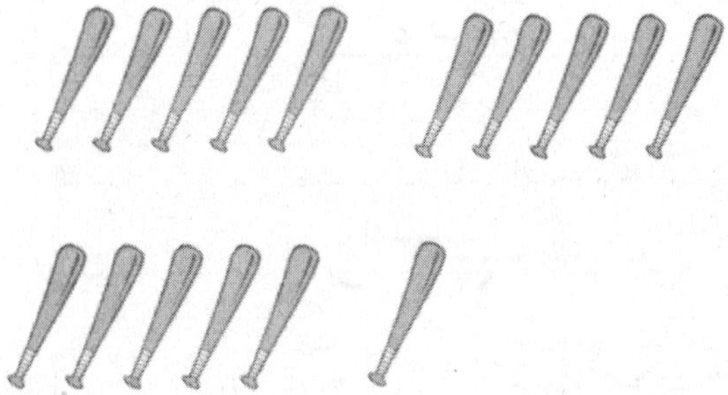

7. 17 - 8 = ____

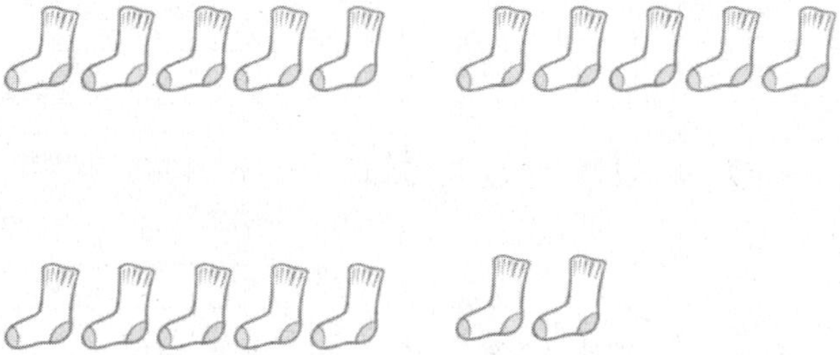

Draw and circle 10, **or** break apart the teen number with a number bond. Then subtract.

8. 12 - 8 = ___

9. 13 - 8 = ___

10. 14 - 8 = ___

11. 15 - 8 = ___

Name ______________________ Date ____________

1. Match the number sentence to the picture or to the number bond.

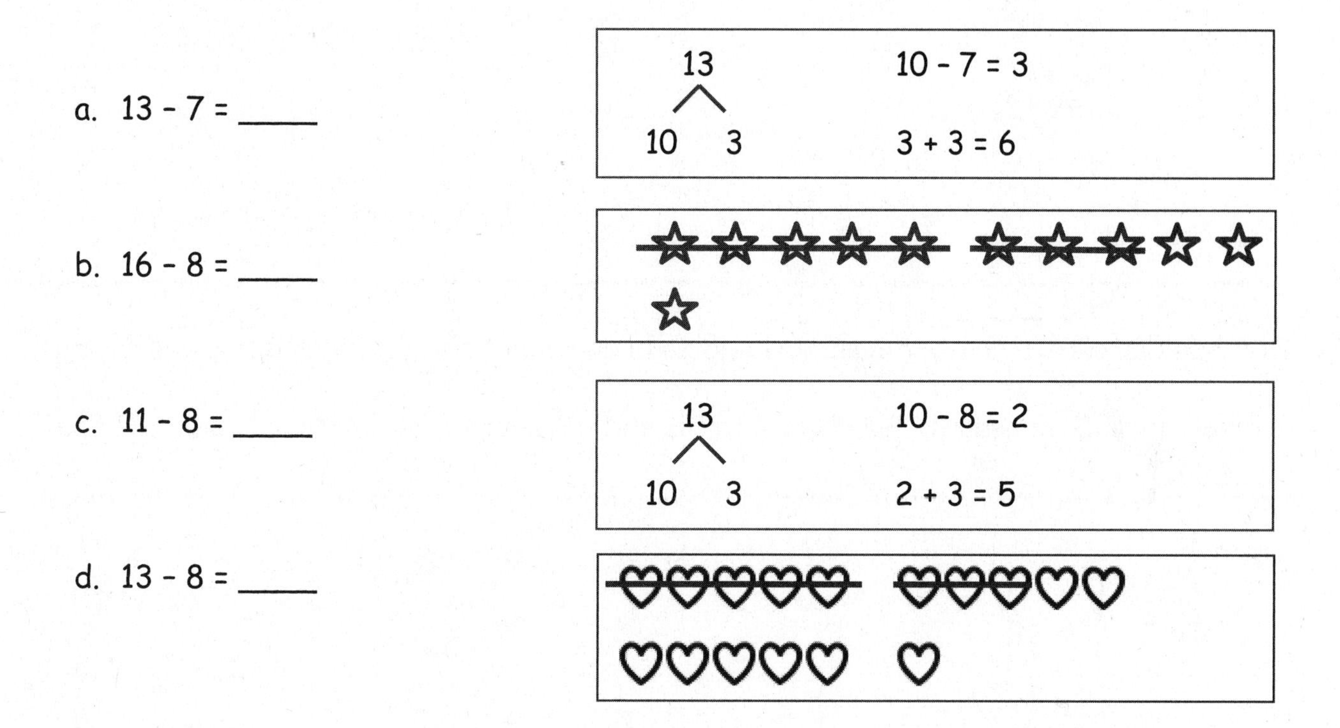

2. Show how you would solve 14 - 8, either with a number bond or a drawing.

Circle 10. Then subtract.

3. Milo has 17 rocks. He throws 8 of them into a pond. How many does he have left?

Milo has _____ rocks left.

Draw and circle 10. Then subtract.

4. Lucy has $12. She spends $8. How much money does she have now?

Lucy has $_____ now.

Draw and circle 10, or use a number bond to break apart the teen number and subtract.

5. Sean has 15 dinosaurs. He gives 8 to his sister. How many dinosaurs does he keep?

Sean keeps _____ dinosaurs.

---

6. Use the picture to fill in the math story. Show a number sentence.

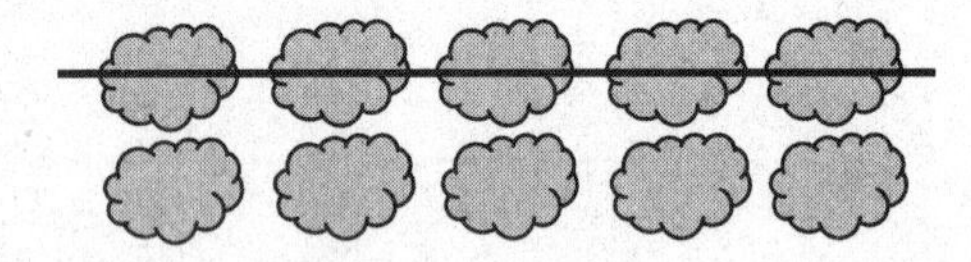

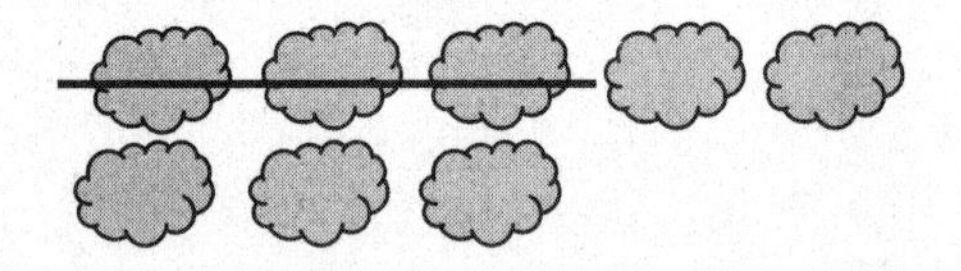

Olivia saw _____ clouds in the sky. _____ clouds went away. How many clouds are left?	Try it! Can you show how to solve this problem with a number bond?

G1-M2-SE-B2-1.3.1-12.2015

Name ______________________________ Date ______________

1. Match the pictures with the number sentences.

a. 13 - 8 = 5	~~ooooo ooo~~oo ooooo oo
b. 14 - 8 = 6	~~ooooo ooo~~oo oooo
c. 17 - 8 = 9	~~ooooo ooo~~oo ooo
d. 18 - 8 = 10	~~ooooo ooo~~oo ooooo o
e. 16 - 8 = 8	~~ooooo ooo~~oo ooooo ooo

Make a math drawing of a 5-group row and some ones to solve the following problems. Write the addition sentence that shows how to add the parts after subtracting 8 or 9.

2. 11 - 8 = _____ ______________________

3. 12 - 8 = _____ ______________________

4. 15 - 8 = _____ ______________________

5. 19 - 8 = ____ ______________________

6. 16 - 8 = ____ ______________________

7. 16 - 9 = ____ ______________________

8. 14 - 9 = ____ ______________________

9. Show how to make ten and take from ten to solve the two number sentences.

a. 6 + 8 = ____

b. 14 - 8 = ____

Name ______________________________ Date ______________

Draw 5-group rows, and cross out to solve. Write the 2+ addition sentence that helped you add the two parts.

1. Annabelle had 13 goldfish. Eight goldfish ate fish food. How many goldfish did not eat fish food?

_____ goldfish did not eat fish food.

2. Sam collected 15 buckets of rain water. He used 8 buckets to water his plants. How many buckets of rain water does Sam have left?

Sam has _____ buckets of rain water left.

3. There were 19 turtles swimming in the pond. Some turtles climbed up onto the dry rocks, and now there are only 8 turtles swimming. How many turtles are on the dry rocks?

There are _____ turtles on the dry rocks.

Show making ten or taking from ten to solve the number sentences.

4. 7 + 8 = ______

5. 15 - 8 = ______

Find the missing number by drawing 5-group rows.

6. 11 - 9 = ______

7. 14 - 9 = ______

8. Draw 5-group rows to show the story. Cross out or use number bonds to solve. Write a number sentence to show how you solved the problem.

There were 14 people at home. Ten people were watching a football game. Four people were playing a board game. Eight people left. How many people stayed?

______ people stayed at home.

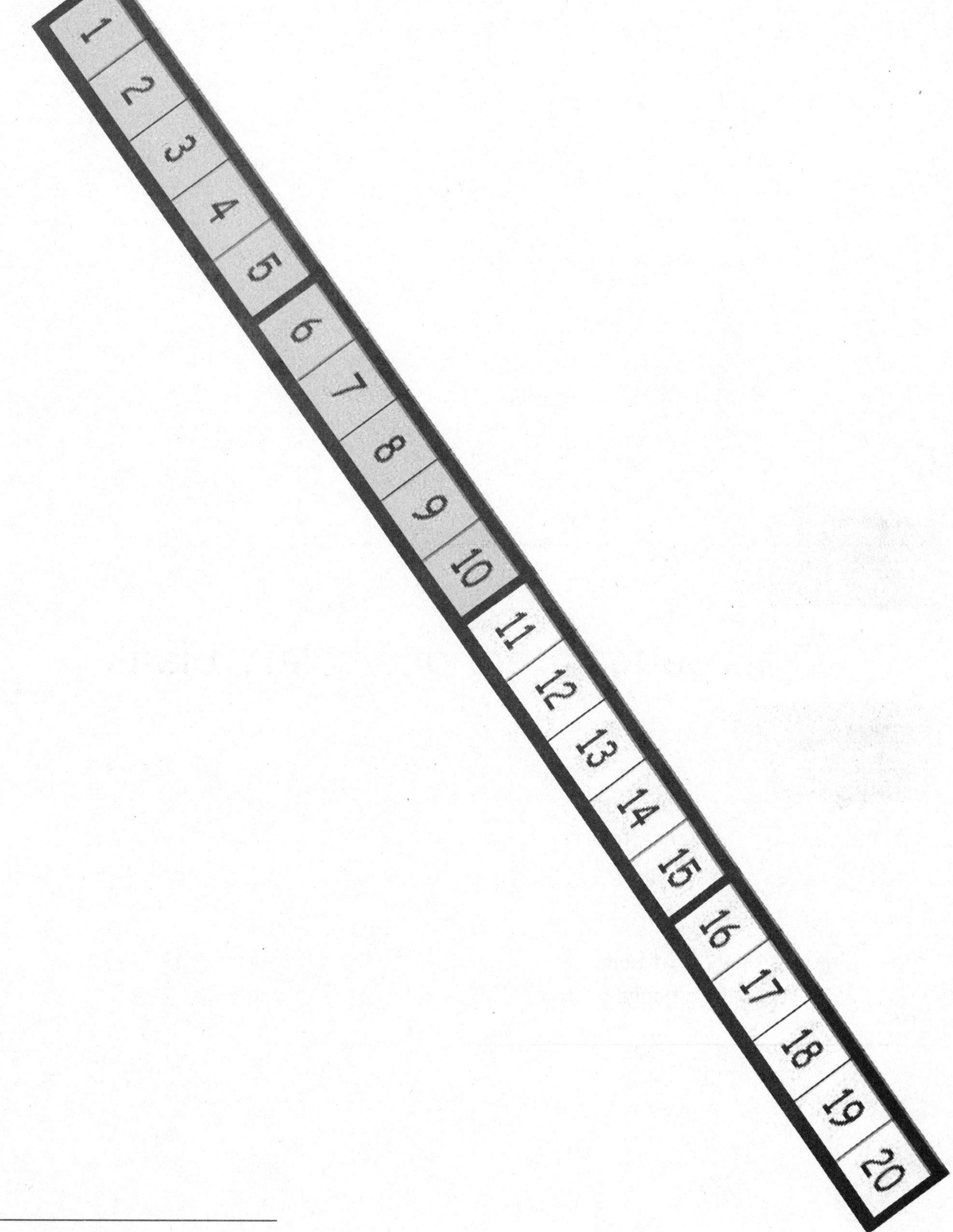

number path 1–20

This page intentionally left blank

Name ______________________________ Date ______________

Use a number bond to show how you used the take from ten strategy to solve the problem.

1. Kevin had 14 crayons. Eight of the crayons were broken. How many of his crayons were not broken?

14 - 8 = ____

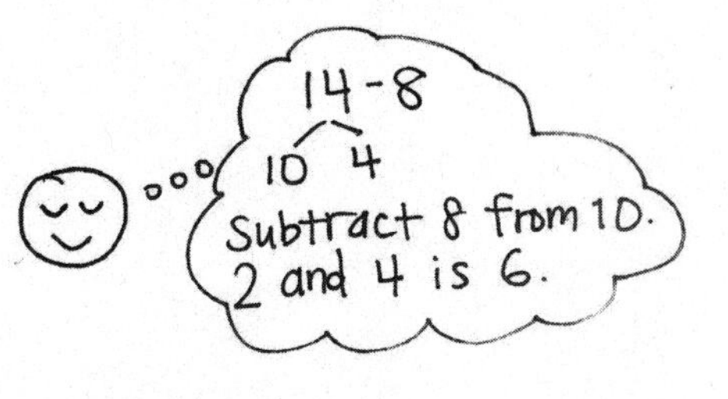

Kevin had ___ crayons that were not broken.

Use number bonds to show your thinking.

2. 17 - 8 = ____

3. 18 - 8 = ____

Count on to solve.

4. 13 - 8 = ____

5. 15 - 8 = ____

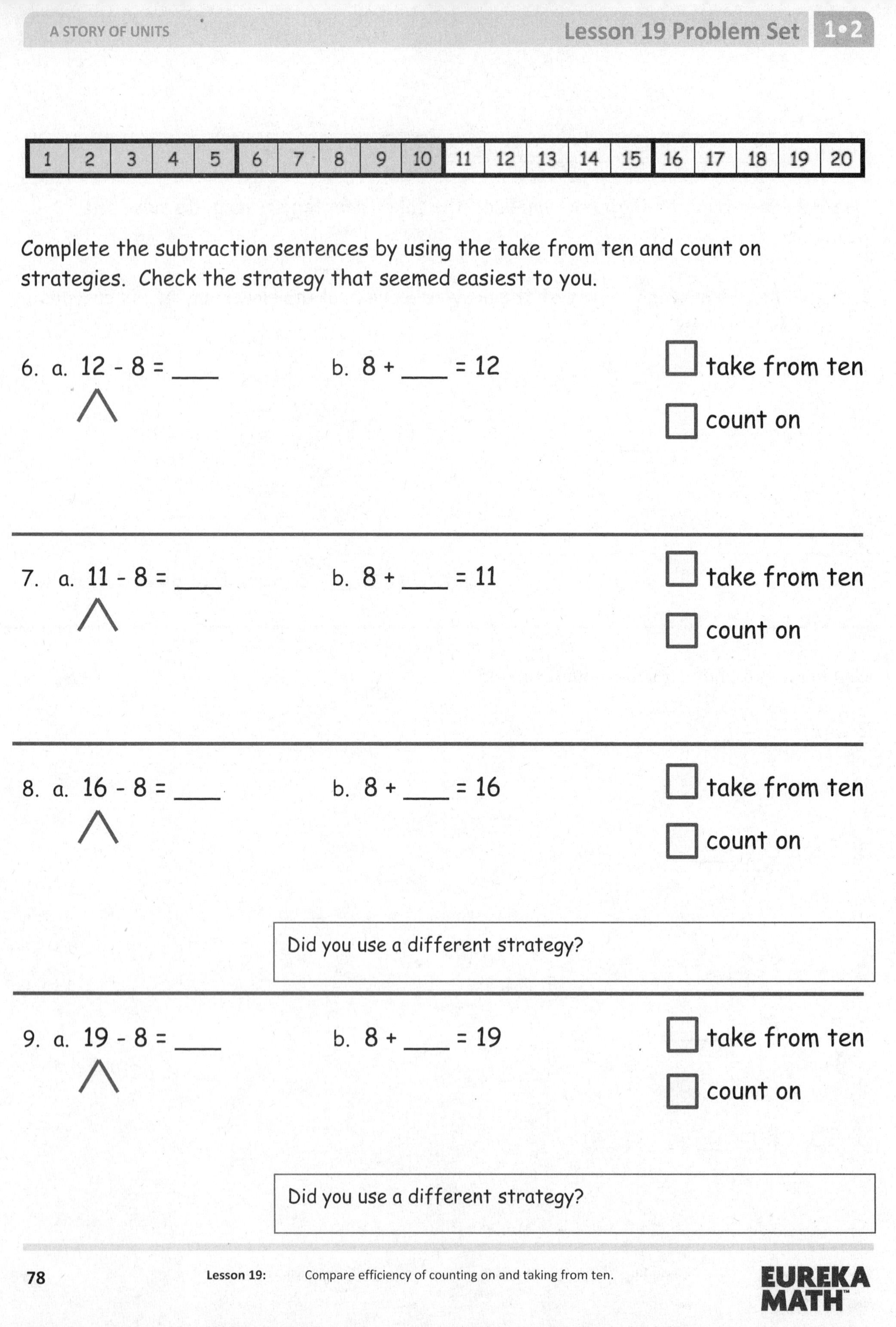

1	2	3	4	5	6	7	8	9	10	11	12	13	14	15	16	17	18	19	20

Complete the subtraction sentences by using the take from ten and count on strategies. Check the strategy that seemed easiest to you.

6. a. 12 - 8 = ____ b. 8 + ____ = 12 ☐ take from ten ☐ count on

7. a. 11 - 8 = ____ b. 8 + ____ = 11 ☐ take from ten ☐ count on

8. a. 16 - 8 = ____ b. 8 + ____ = 16 ☐ take from ten ☐ count on

Did you use a different strategy?

9. a. 19 - 8 = ____ b. 8 + ____ = 19 ☐ take from ten ☐ count on

Did you use a different strategy?

Name ______________________ Date __________

Complete the subtraction sentences by using the take from ten strategy and count on.

1	2	3	4	5	6	7	8	9	10	11	12	13	14	15	16	17	18	19	20

1. a. 12 - 8 = ____ b. 8 + ____ = 12

2. a. 15 - 8 = ____ b. 8 + ____ = 15

Choose the count on strategy or the take from ten strategy to solve.

3. 11 - 8 = ____

4. 17 - 8 = ____

Use a number bond to show how you solved using the take from ten strategy.

5. Elise counted 16 worms on the pavement. Eight worms crawled into the dirt. How many worms did Elise still see on the pavement?

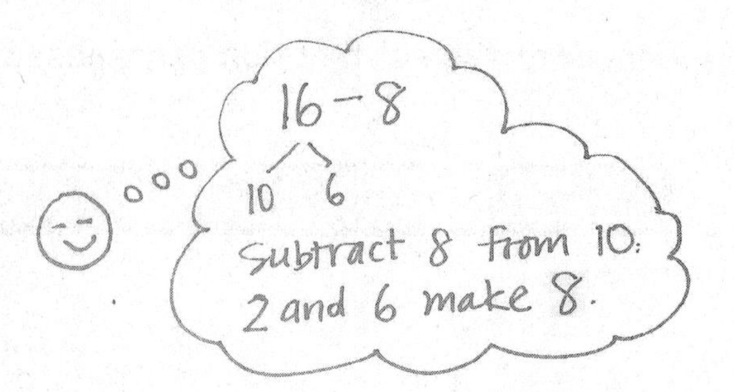

16 - 8 = _____

Elise still saw ____ worms on the pavement.

---

6. John ate 8 orange slices. If he started with 13, how many orange slices does he have left?

John has ____ orange slices left.

---

7. Match the addition number sentence to the subtraction number sentence. Fill in the missing numbers.

a. 12 - 8 = _____ | 8 + _____ = 11

b. 15 - 8 = _____ | 8 + _____ = 18

c. 18 - 8 = _____ | 8 + _____ = 12

d. 11 - 8 = _____ | 8 + _____ = 15

Name ____________________ Date __________

Complete the number sentences to make them true.

1. 15 - 9 = ____
2. 15 - 8 = ____
3. 15 - 7 = ____

4. 17 - 9 = ____
5. 17 - 8 = ____
6. 17 - 7 = ____

7. 16 - 9 = ____
8. 16 - 8 = ____
9. 16 - 7 = ____

10. 19 - 9 = ____
11. 19 - 8 = ____
12. 19 - 7 = ____

13. Match equal expressions.

a. 19 - 9 12 - 7

b. 13 - 8 18 - 8

14. Read the math story. Use a drawing or a number bond to show how you know who is right.

  a. Elsie says that the expressions 17 - 8 and 18 - 9 are equal. John says they are not equal. Who is right?

  b. John says that the expressions 11 - 8 and 12 - 8 are not equal. Elsie says they are. Who is right?

  c. Elsie says that to solve 17 - 9, she can take one from 17 and give it to 9 to make 10. So, 17 - 9 is equal to 16 - 10. John thinks Elsie made a mistake. Who is correct?

  d. John and Elsie are trying to find several subtraction number sentences that start with numbers larger than 10 and have an answer of 7. Help them figure out number sentences. They started the first one.

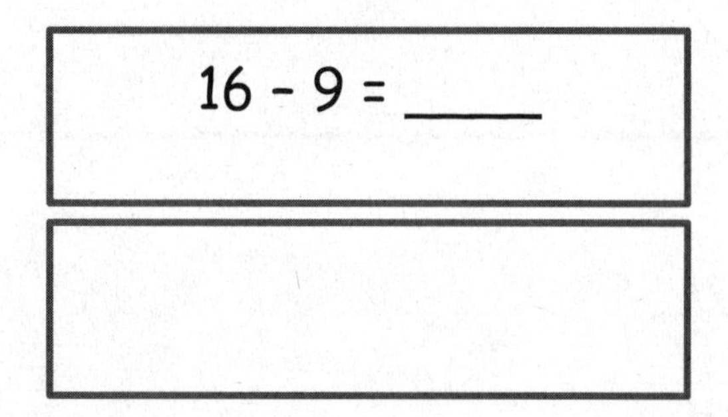

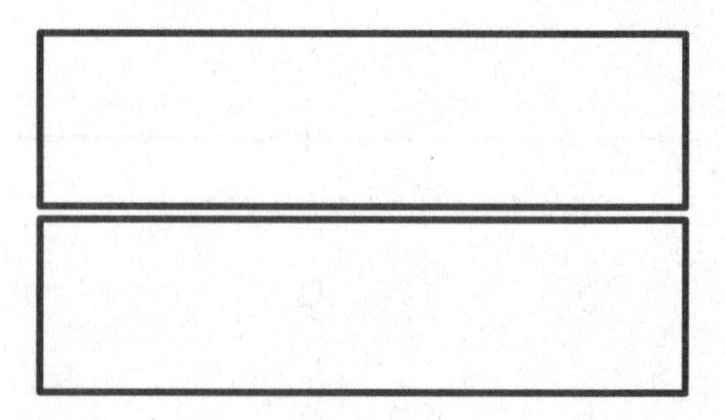

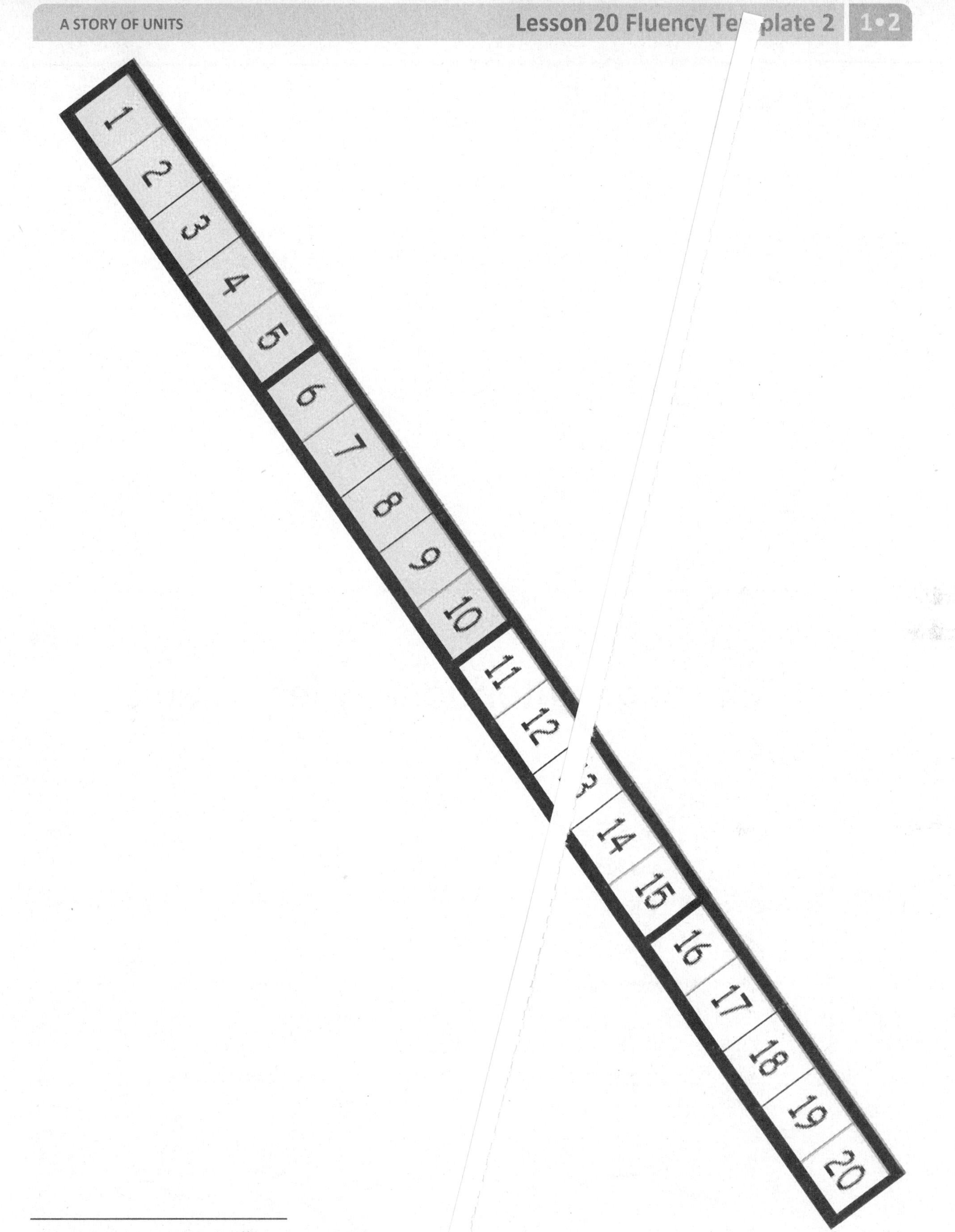

number path 1–20; originally in Lesson 18

This page intentionally left blank

Name ____________________ Date ____________

Olivia and Jake both solved the word problems.
Write the strategy used under their work.
Check their work. If incorrect, solve correctly.
If solved correctly, solve using a different strategy.

Strategies:

- Take from 10
- Make 10
- Count on
- I just knew

1. A fruit bowl had 13 apples. Mike ate 6 apples from the fruit bowl. How many apples were left?

Olivia's work

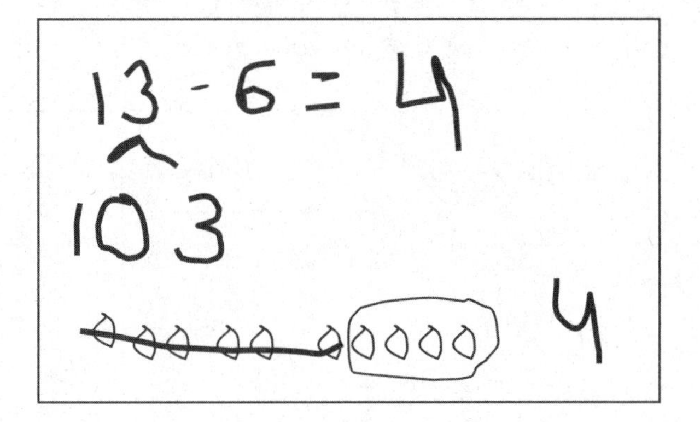

Jake's work

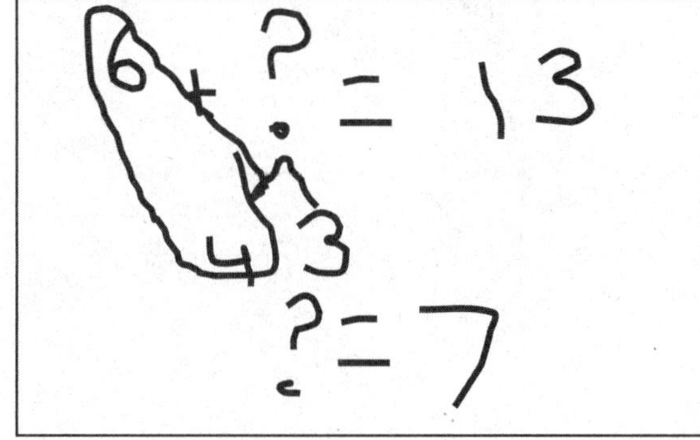

a. Strategy: ____________________

b. Strategy: ____________________

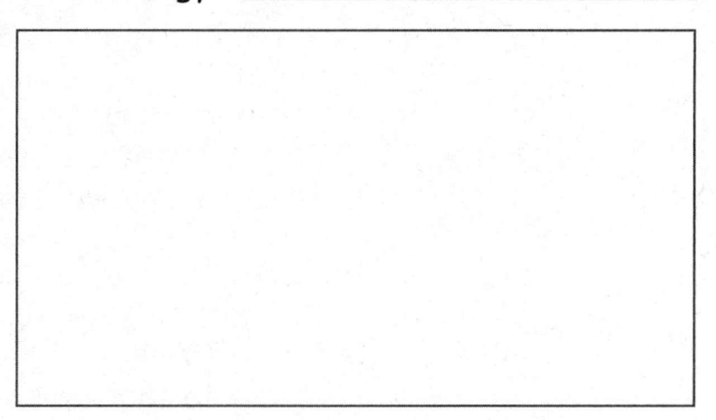

c. Explain your strategy choice below.

2. Drew has 17 baseball cards in a box. He has 8 cards with Red Sox players, and the rest are Yankees players. How many Yankees player cards does Drew have in his box?

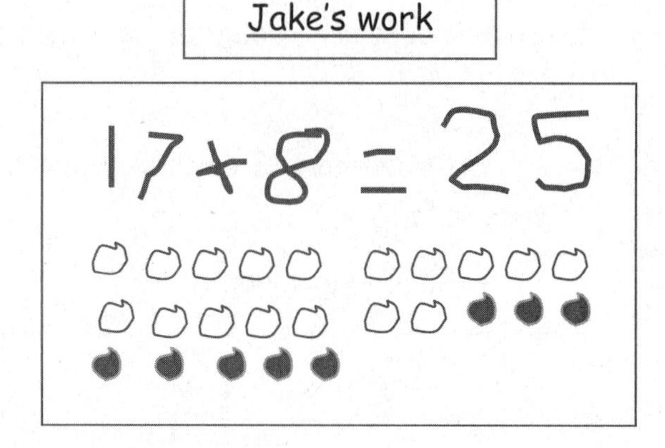

a. Strategy: ____________________

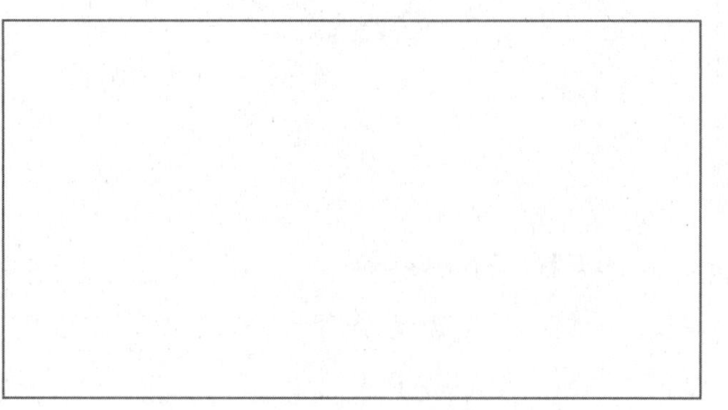

b. Strategy: ____________________

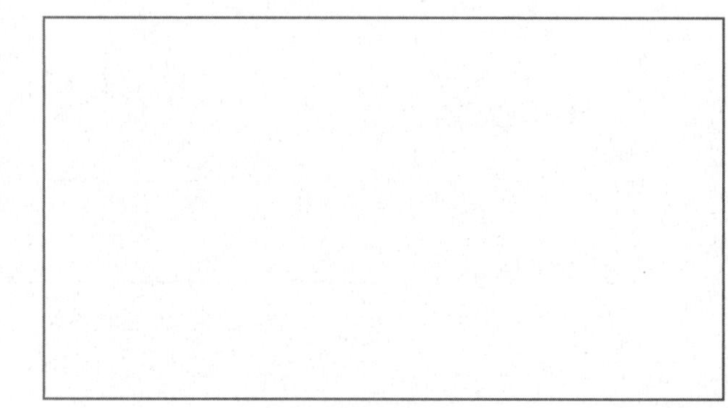

c. Explain your strategy choice below.

Name ______________________________ Date ______________

Read the word problem.
Draw and label.
Write a number sentence and a statement that matches the story.

1. This week, Maria ate 5 yellow plums and some red plums. If she ate 11 plums in all, how many red plums did Maria eat?

---

2. Tatyana counted 14 frogs. She counted 8 swimming in the pond and the rest sitting on lily pads. How many frogs did she count sitting on lily pads?

3. Some children are on the playground. Eight are on the swings, and the rest are playing tag. There are 15 children in all. How many children are playing tag?

4. Oziah read some non-fiction books. Then, he read 7 fiction books. If he read 16 books altogether, how many non-fiction books did Oziah read?

Meet with a partner, and share your drawings and sentences.
Talk with your partner about how your drawing matches the story.

Name ______________________________ Date ______________

Read the word problem.
Draw and label.
Write a number sentence and a statement that matches the story.

Remember to draw a box around your solution in the number sentence.

Strategies:

- Take from 10
- Make 10
- Count on
- I just knew

1. Michael and Anastasia pick 14 flowers for their mom. Michael picks 6 flowers. How many flowers does Anastasia pick?

2. Daquan bought 6 toy cars. He also bought some magazines. He bought 15 items in all. How many magazines did Daquan buy?

3. Henry and Millie baked 18 cookies. Nine of the cookies were chocolate chip. The rest were oatmeal. How many were oatmeal?

4. Felix made 8 birthday invitations with hearts. He made the rest with stars. He made 17 invitations in all. How many invitations had stars?

5. Ben and Miguel are having a bowling contest. Ben wins 9 times. They play 17 games in all. There are no tied games. How many times does Miguel win?

6. Kenzie went to soccer practice 16 days this month. Only 9 of her practices were on a school day. How many times did she practice on a weekend?

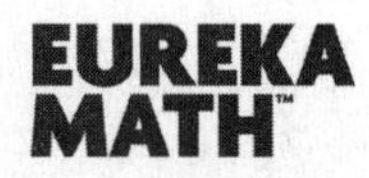

Name ______________________________ Date ______________

Read the word problem.
Draw and label.
Write a number sentence and a statement that matches the story.

1. Janet read 8 books during the week. She read some more books on the weekend. She read 12 books total. How many books did Janet read on the weekend?

2. Eric scored 13 goals this season! He scored 5 goals before the playoffs. How many goals did Eric score during the playoffs?

3. There were 8 ladybugs on a branch. Some more came. Then, there were 15 ladybugs on the branch. How many ladybugs came?

---

4. Marco's friend gave him some baseball cards at school. If he was already given 9 baseball cards by his family, and he now has 19 cards in all, how many baseball cards did he get in school?

Meet with a partner and share your drawings and sentences. Talk with your partner about how your drawing matches the story.

Name ______________________________ Date ______________

Read the word problem.
Draw and label.
Write a number sentence and a statement that matches the story.

1. Micah collected 9 pinecones on Friday and some more on Saturday. Micah collected a total of 14 pinecones. How many pinecones did Micah collect on Saturday?

---

2. Giana bought 8 star stickers to add to her collection. Now, she has 17 stickers in all. How many stickers did Giana have at first?

3. Samil counted 5 pigeons on the street. Some more pigeons came. There were 13 pigeons in all. How many pigeons came?

---

4. Claire had some eggs in the fridge. She bought 12 more eggs. Now, she has 18 eggs in all. How many eggs did Claire have in the fridge at first?

Name ______________________________ Date ______________

Read the word problem.
Draw and label.
Write a number sentence and a statement that match the story.

1. Jose sees 11 frogs on the shore. Some of the frogs hop into the water. Now, there are 8 frogs on the shore. How many frogs hopped into the water?

2. Cameron gives some of his apples to his sister. He still has 9 apples left. If he had 15 apples at first, how many apples did he give to his sister?

3. Molly had 16 books. She loaned some to Gia. How many books did Gia borrow if Molly has 8 books left?

---

4. Eighteen baby goats were playing outside. Some went into the barn. Nine stayed outside to play. How many baby goats went inside?

Meet with a partner and share your drawings and sentences. Talk with your partner about how your drawing tells the story.

Name ______________________________ Date ______________

Read the word problem.
Draw and label.
Write a number sentence and a statement that matches the story.

1. Toby dropped 12 crayons on the classroom floor. Toby picked up 9 crayons. Marnie picked up the rest. How many crayons did Marnie pick up?

---

2. There were 11 students on the playground. Some students went back into the classroom. If 7 students stayed outside, how many students went inside?

3. At the play, 8 students from Mr. Frank's room got a seat. If there were 17 children from Room 24, how many children did not get a seat?

---

4. Simone had 12 bagels. She shared some with friends. Now, she has 9 bagels left. How many did she share with friends?

Name ______________________________ Date ______________

Use the expression cards to play Memory. Write the matching expressions to make true number sentences.

1.

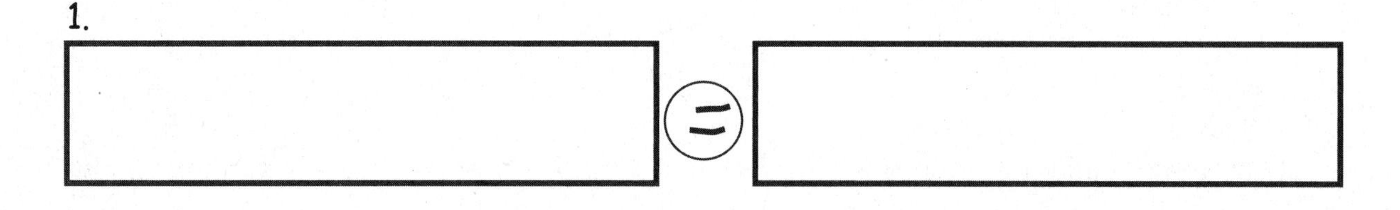

2.

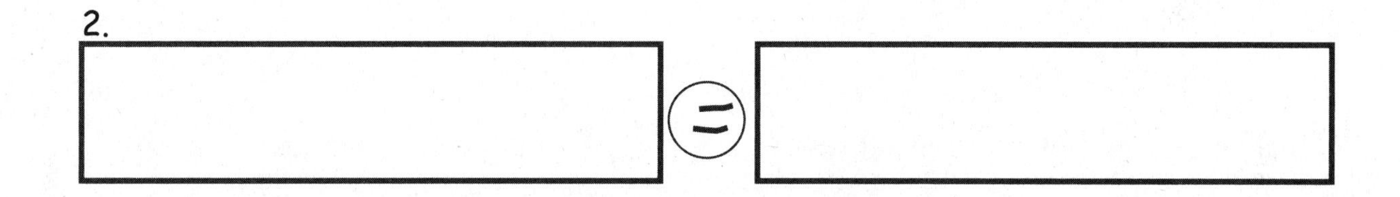

3.

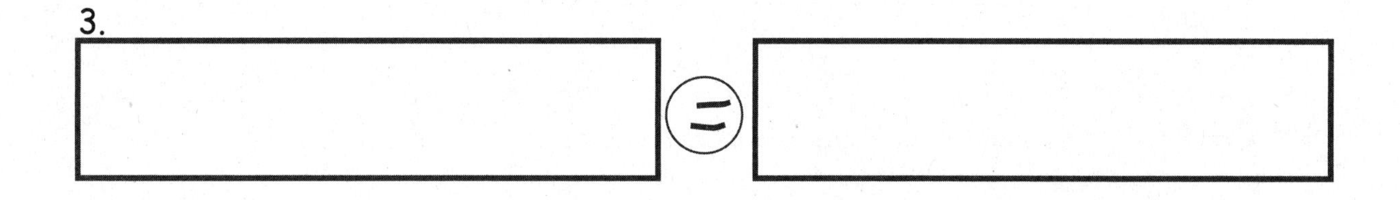

4.

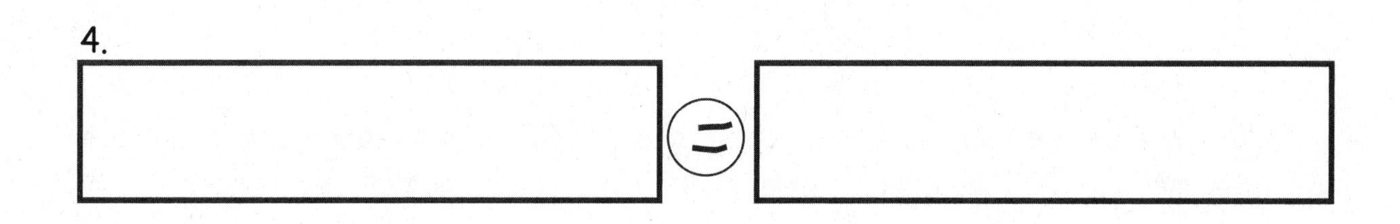

5.

6. Write a true number sentence using the expressions that you have left over. Use pictures and words to show how you know two of the expressions have the same unknown numbers.

7. Use other facts you know to write at least two true number sentences similar to the type above.

8. The following addition number sentences are FALSE. Change one number in each problem to make a TRUE number sentence, and rewrite the number sentence.

a. 8 + 5 = 10 + 2 ____________________

b. 9 + 3 = 8 + 5 ____________________

c. 10 + 3 = 7 + 5 ____________________

9. The following subtraction number sentences are FALSE. Change one number in each problem to make a TRUE number sentence, and rewrite the number sentence.

a. 12 - 8 = 1 + 2 ____________________

b. 13 - 9 = 1 + 4 ____________________

c. 1 + 3 = 14 - 9 ____________________

b. Charlie picked 11 - 4, and Lola picked 6 + 1. Charlie says these expressions are not equal, but Lola disagrees. Who is right? Use a picture to explain your thinking.

c. Lola picked 9 + 7, and Charlie picked 15 - 8. Lola says these expressions are equal but Charlie disagrees. Who is right? Use a picture to explain your thinking.

3. The following addition number sentences are FALSE. Change one number in each problem to make a TRUE number sentence, and rewrite the number sentence.

a. 10 + 5 = 9 + 5 ____________________

b. 10 + 3 = 8 + 4 ____________________

c. 9 + 3 = 8 + 5 ____________________

Name ______________________________ Date ____________

1. Circle "true" or "false."

Equation	True or False?
a. 2 + 3 = 5 + 1	True / False
b. 7 + 9 = 6 + 10	True / False
c. 11 - 8 = 12 - 9	True / False
d. 15 - 4 = 14 - 5	True / False
e. 18 - 6 = 2 + 10	True / False
f. 15 - 8 = 2 + 5	True / False

2. Lola and Charlie are using expression cards to make true number sentences. Use pictures and words to show who is right.

   a. Lola picked 4 + 8, and Charlie picked 9 + 3. Lola says these expressions are eq but Charlie disagrees. Who is right? Explain your thinking.

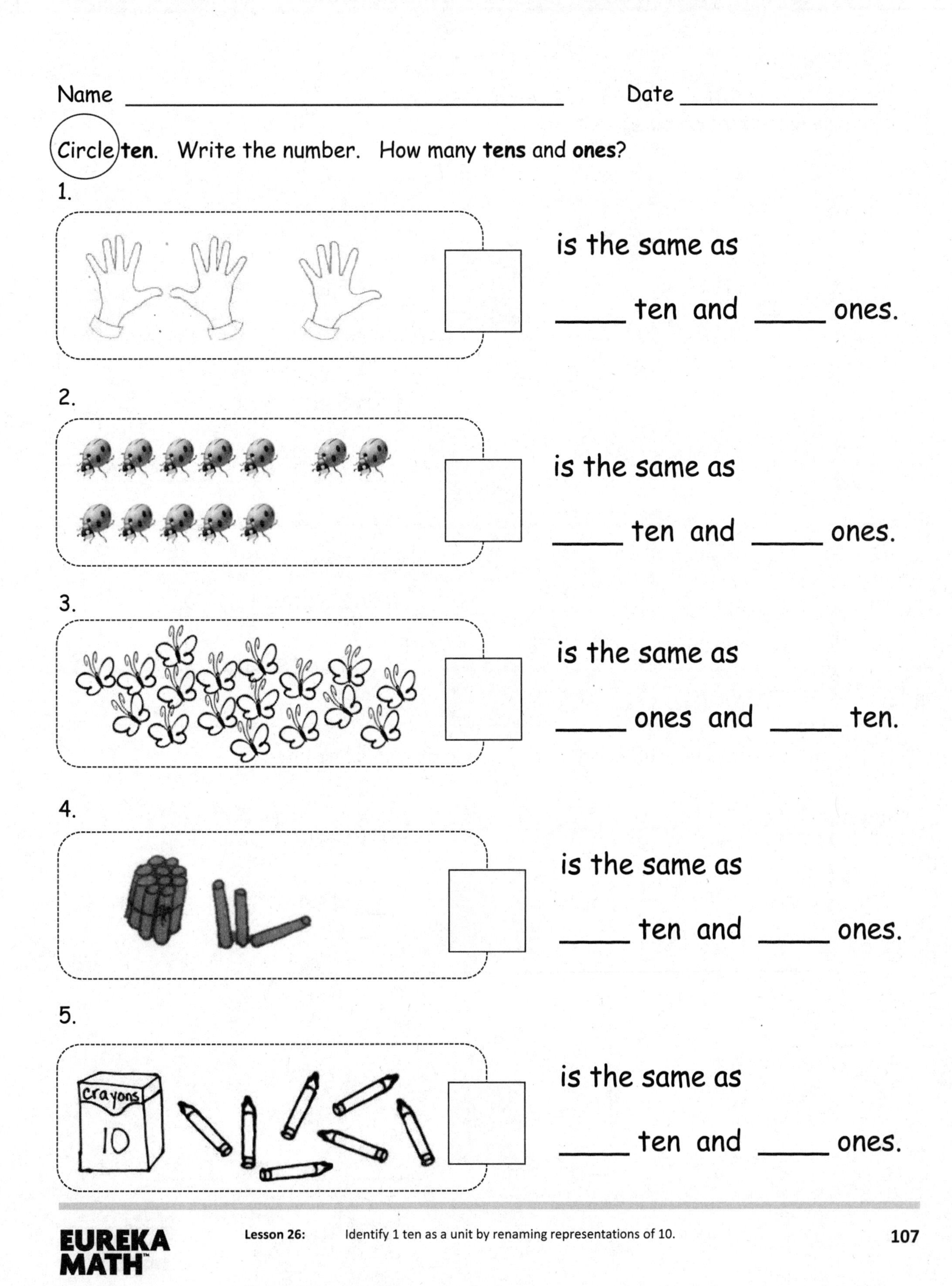

Name ______________________ Date ____________

Circle **ten**. Write the number. How many **tens** and **ones**?

1.

☐ is the same as

____ ten and ____ ones.

2.

☐ is the same as

____ ten and ____ ones.

3.

☐ is the same as

____ ones and ____ ten.

4.

☐ is the same as

____ ten and ____ ones.

5.

☐ is the same as

____ ten and ____ ones.

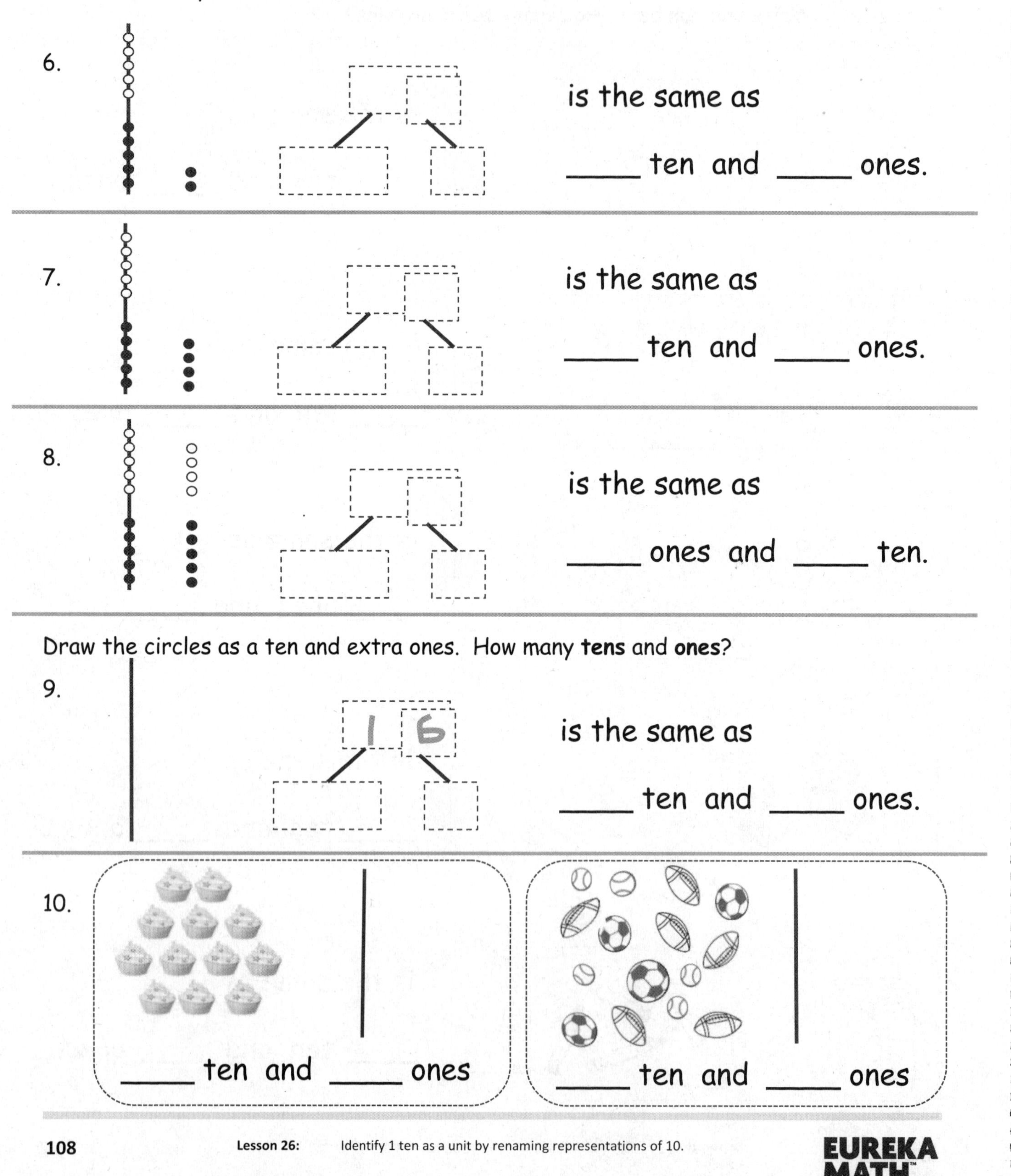

Show the total and tens and ones with Hide Zero cards.
Write how many **tens** and **ones**.

6. is the same as

____ ten and ____ ones.

7. is the same as

____ ten and ____ ones.

8. is the same as

____ ones and ____ ten.

Draw the circles as a ten and extra ones. How many **tens** and **ones**?

9. is the same as

____ ten and ____ ones.

10. ____ ten and ____ ones

____ ten and ____ ones

EUREKA MATH™

Name ______________________________ Date ______________

Circle **ten**. Write the number. How many **tens** and **ones**?

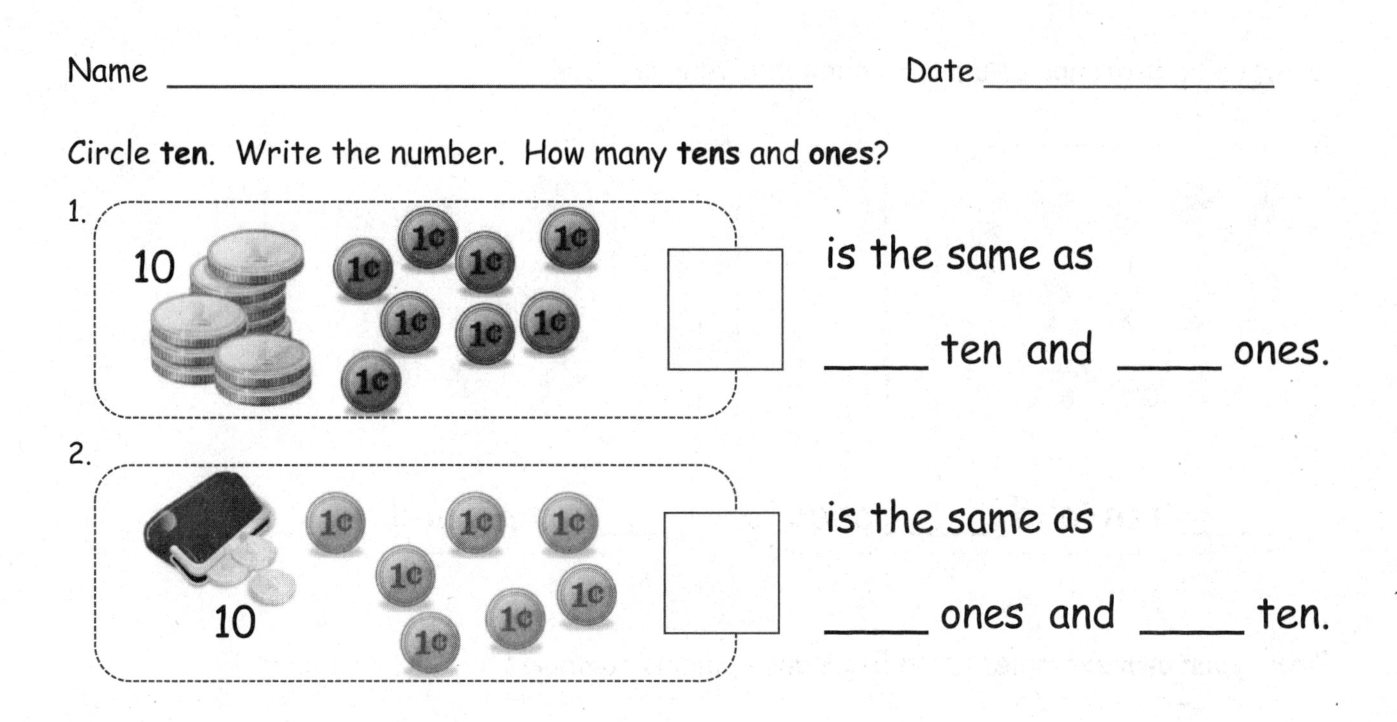

1. [ ] is the same as

____ ten and ____ ones.

2. [ ] is the same as

____ ones and ____ ten.

Use the Hide Zero pictures to draw the ten and ones shown on the cards.

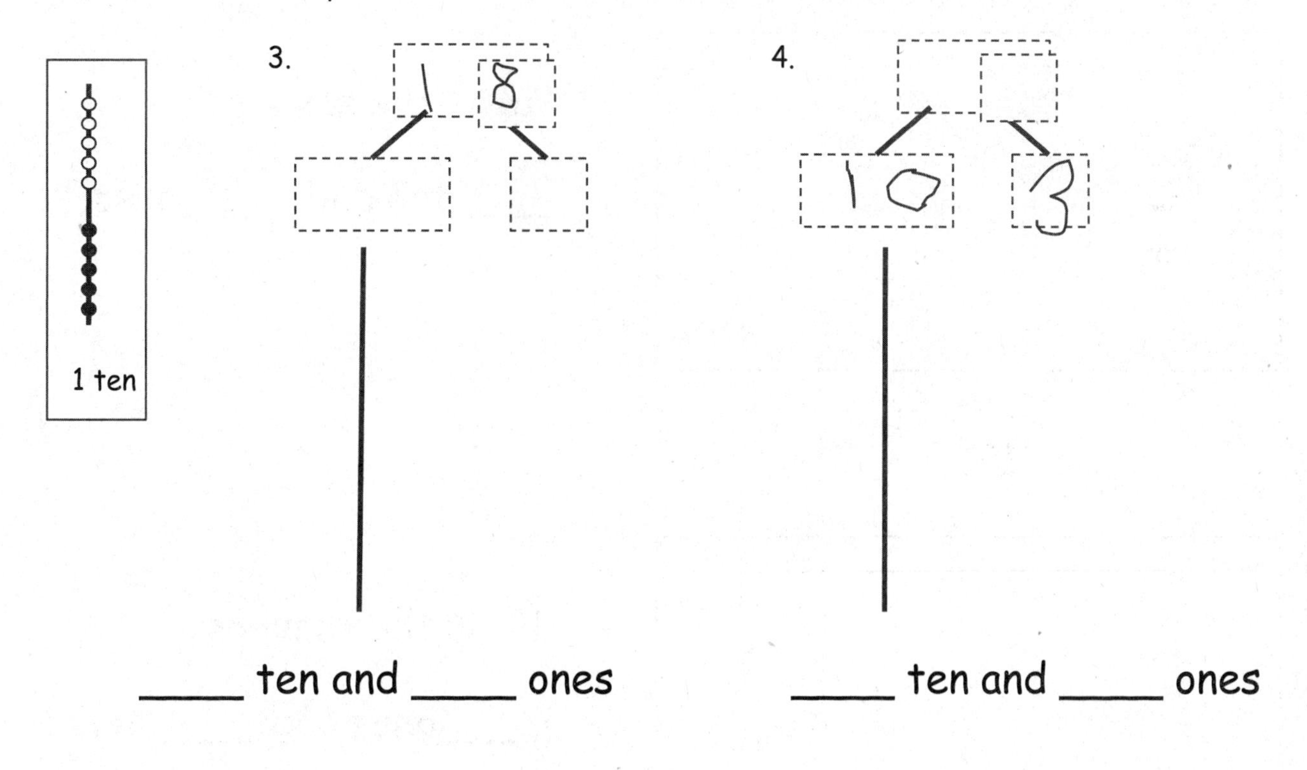

____ ten and ____ ones ____ ten and ____ ones

Draw using 5-groups columns to show the tens and ones.

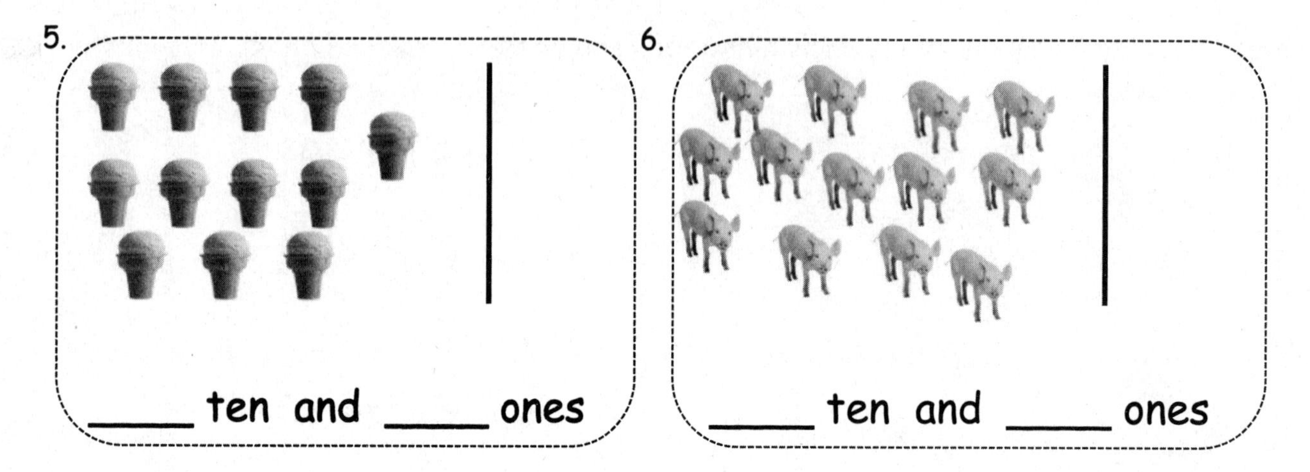

Draw your own examples using 5-groups columns to show the tens and ones.

7. 16

16 is the same as

___ ten and ____ ones.

8. 19

19 is the same as

____ ones and ____ ten.

Name ______________________________ Date ________________

Solve the problems. Write your answers to show how many **tens** and **ones**. If there is only 1 ten, cross off the "s."

Add.

1. 12 + 6 = [ ][ ]

____ tens and ____ ones

2. 5 + 13 = [ ][ ]

____ tens and ____ ones

3. 8 + 7 = [ ][ ]

____ tens and ____ ones

4. [ ][ ] = 8 + 12

____ tens and ____ ones

Subtract.

5. 17 - 4 = [ ][ ]

____ tens and ____ ones

6. 17 - 5 = [ ][ ]

____ tens and ____ ones

7. 14 - 6 = [ ][ ]

____ tens and ____ ones

8. [ ][ ] = 16 - 7

____ tens and ____ ones

Read the word problem. Draw and label. Write a number sentence and statement that matches the story. Rewrite your answer to show its tens and ones. If there is only 1 ten or 1 one, cross off the "s."

9. Frankie and Maya made 4 big sandcastles at the beach. If they made 10 small sandcastles, how many total sandcastles did they make?

____ tens and ____ ones

10. Ronnie has 8 stickers that are stars. Her friend Sina gives her 7 more. How many stickers does Ronnie have now?

____ tens and ____ ones

11. We tied 14 balloons to the tables for a party, but 3 floated away! How many balloons were still tied to the tables?

____ tens and ____ ones

12. I ate 5 of the 16 strawberries that I picked. How many did I have left over?

____ tens and ____ ones

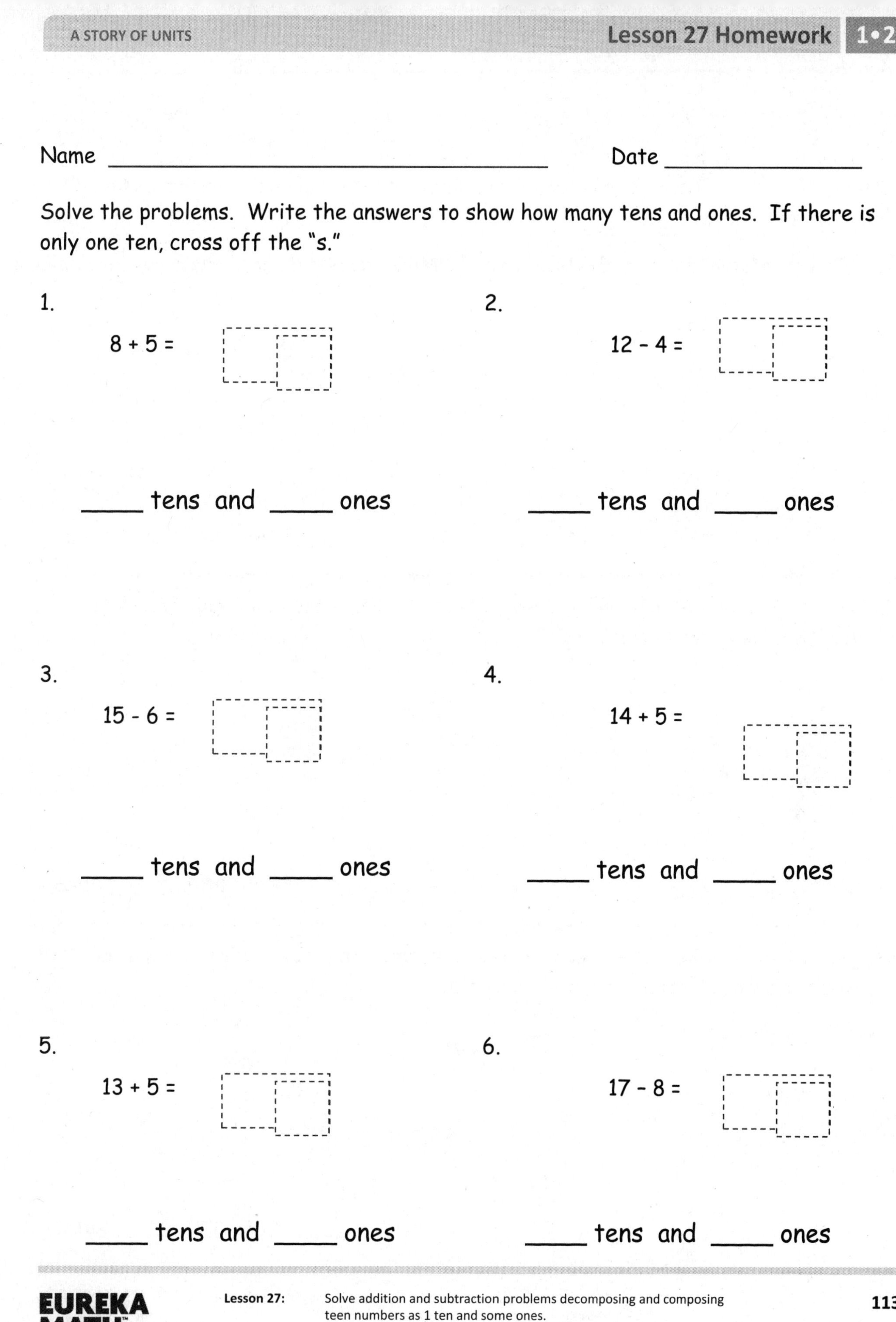

Name ____________________ Date ____________

Solve the problems. Write the answers to show how many tens and ones. If there is only one ten, cross off the "s."

1. 8 + 5 =

____ tens and ____ ones

2. 12 - 4 =

____ tens and ____ ones

3. 15 - 6 =

____ tens and ____ ones

4. 14 + 5 =

____ tens and ____ ones

5. 13 + 5 =

____ tens and ____ ones

6. 17 - 8 =

____ tens and ____ ones

Read the word problem. Draw and label. Write a number sentence and statement that matches the story. Rewrite your answer to show its tens and ones. If there is only 1 ten, cross off the "s."

7. Mike has some red cars and 8 blue cars. If Mike has 9 red cars, how many cars does he have in all?

_____ tens and _____ ones

8. Yani and Han had 14 golf balls. They lost some balls. They had 8 golf balls left. How many balls did they lose?

_____ tens and _____ ones

9. Nick rides his bike for 6 miles over the weekend. He rides 14 miles during the week. How many total miles does Nick ride?

_____ tens and _____ ones

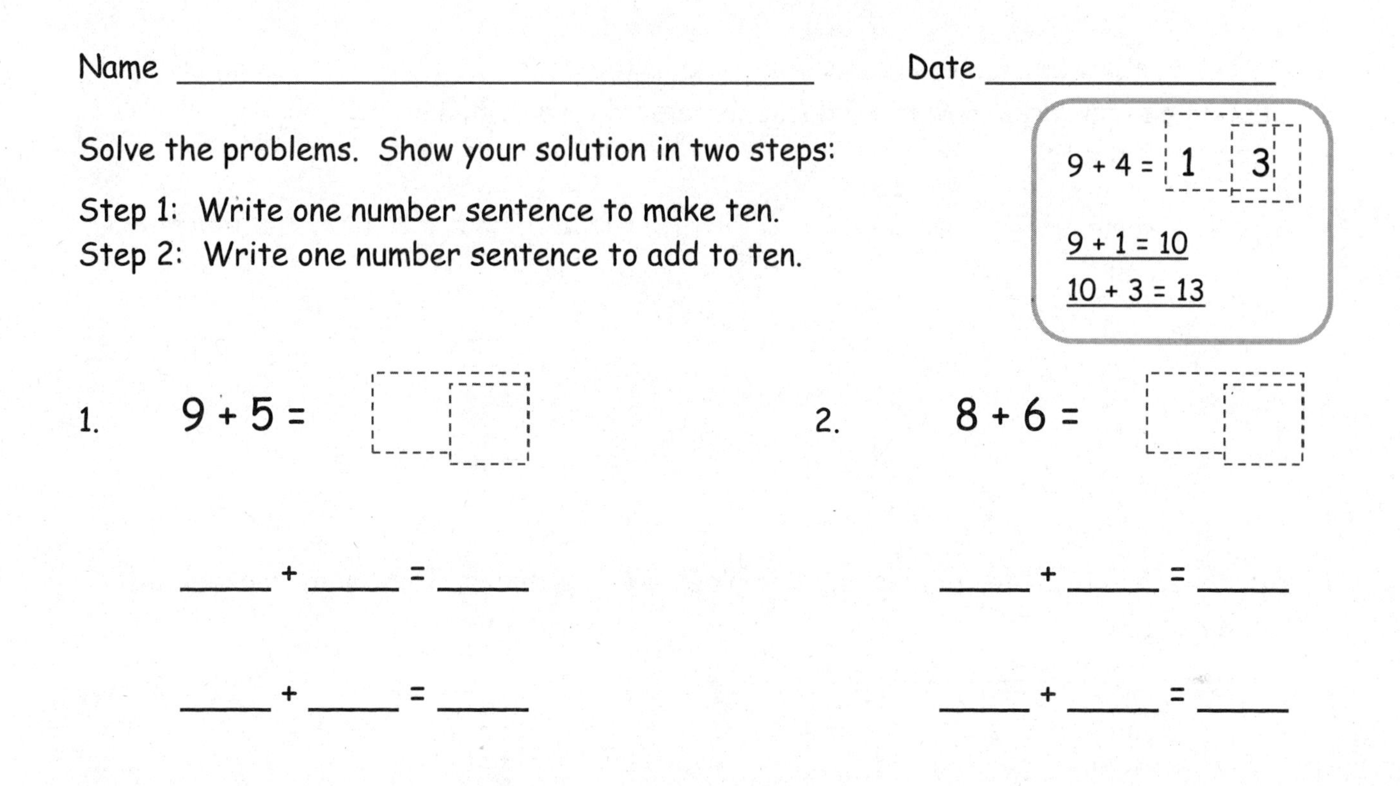

Name ______________________ Date ____________

Solve the problems. Show your solution in two steps:

Step 1: Write one number sentence to make ten.
Step 2: Write one number sentence to add to ten.

9 + 4 = 1 3

9 + 1 = 10
10 + 3 = 13

1. 9 + 5 =

____ + ____ = ____

____ + ____ = ____

2. 8 + 6 =

____ + ____ = ____

____ + ____ = ____

Solve. Then, write a statement to show your answer.

3. Su-Hean put together a collage with 9 pictures. Adele put together another collage with 6 pictures. How many pictures did they use?

9 + 6 = ____

____ + ____ = ____

____ + ____ = ____

4. Imran has 8 crayons in his pencil case and 7 crayons in his desk. How many crayons does Imran have altogether?

____ + ____ = ____

____ + ____ = ____

5. At the park, there were 4 ducks swimming in the pond. If there were 9 ducks resting on the grass, how many ducks were at the park in all?

6. Cece made 7 frosted cookies and 8 cookies with sprinkles. How many cookies did Cece make?

7. Payton read 8 books about dolphins and whales. She read 9 books about dogs and cats. How many books did she read about animals altogether?

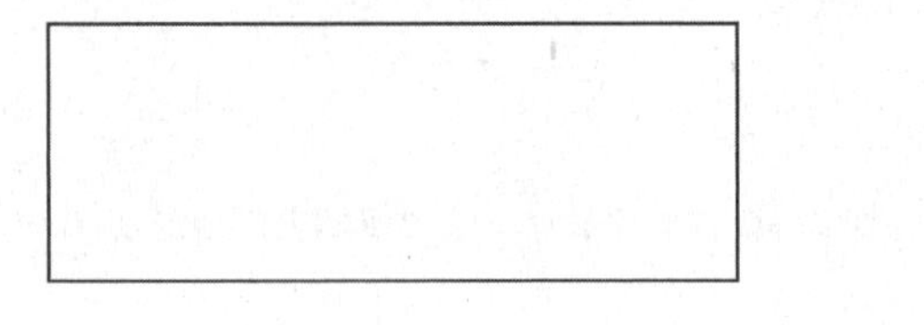

Name ______________________ Date ____________

Solve the problems. Write your answers to show how many **tens** and **ones**.

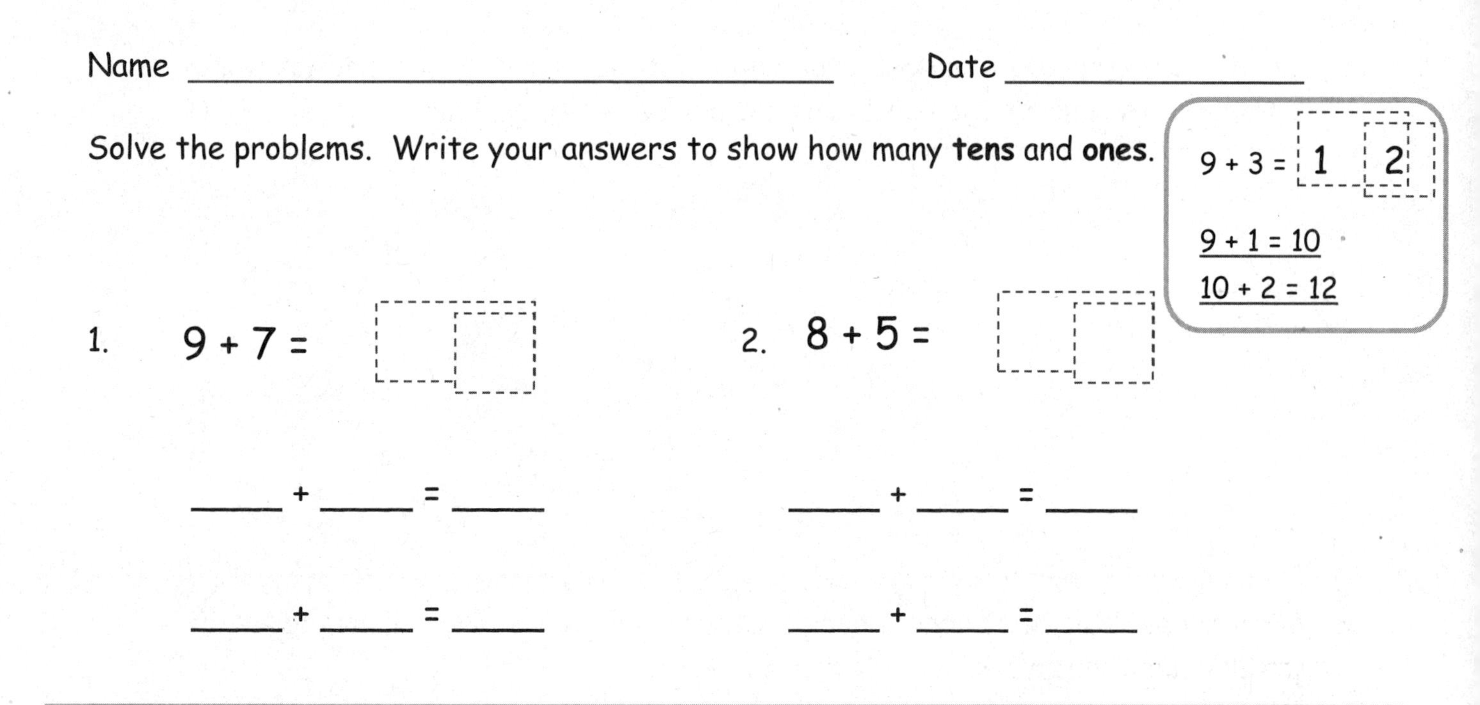

Solve. Write the two number sentences for each step to show how you make **a ten**.

3. Boris has 9 board games on his shelf and 8 board games in his closet. How many board games does Boris have altogether?

4. Sabra built a tower with 8 blocks. Yuri put together another tower with 7 blocks. How many blocks did they use?

5. Camden solved 6 addition word problems. She also solved 9 subtraction word problems. How many word problems did she solve altogether?

6. Minna made 4 bracelets and 8 necklaces with her beads. How many pieces of jewelry did Minna make?

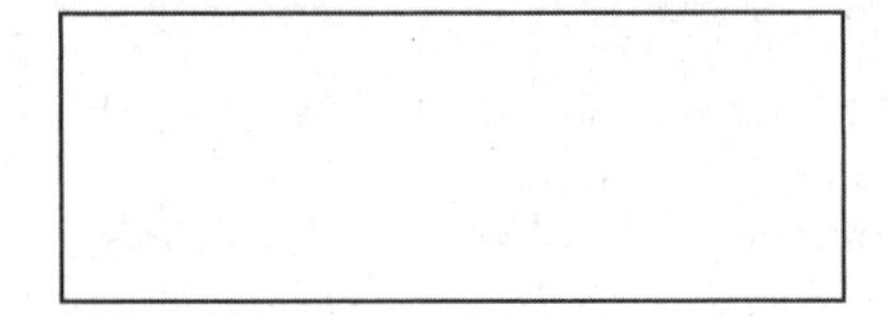

7. I put 5 peaches into my bag at the farmer's market. If I already had 7 apples in my bag, how many pieces of fruit did I have in all?

Name ______________________ Date ______________

Solve the problems. Write your answers to show how many **tens** and **ones**. Show your solution in two steps:

Step 1: Write one number sentence to subtract from ten.

Step 2: Write one number sentence to add the remaining parts.

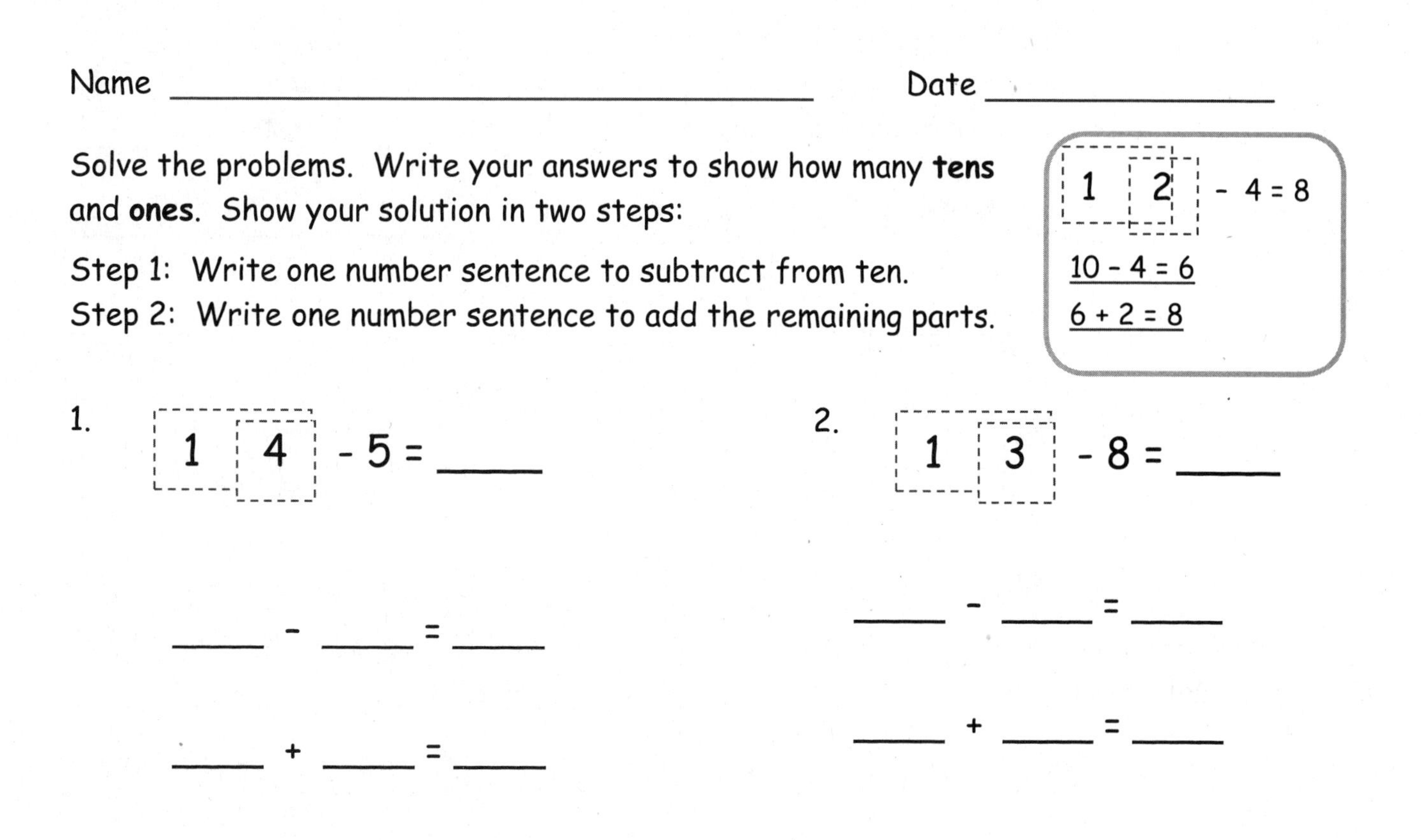

3. Tatyana counted 14 frogs. She counted 8 swimming in the pond and the rest sitting on lily pads. How many frogs did she count sitting on lily pads?

4. This week, Maria ate 5 yellow plums and some red plums. If she ate 11 plums in all, how many red plums did Maria eat?

5. Some children are on the playground playing tag. Eight are on the swings. If there are 16 children on the playground in all, how many children are playing tag?

6. Oziah read some nonfiction books. Then, he read 6 fiction books. If he read 18 books altogether, how many nonfiction books did Oziah read?

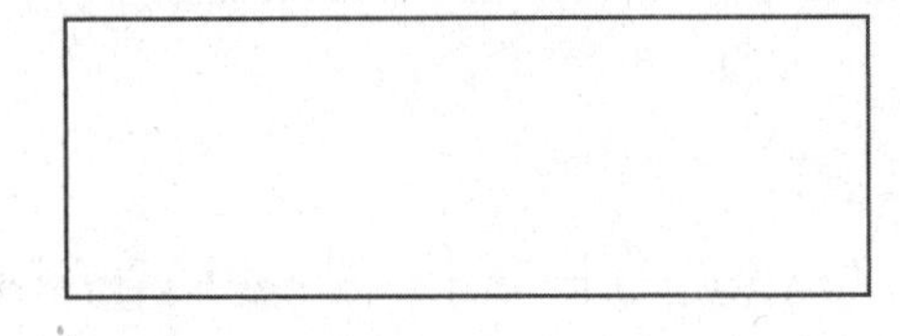

7. Hadley has 9 buttons on her jacket. She has some more buttons on her shirt. Hadley has a total of 17 buttons on her jacket and shirt. How many buttons does she have on her shirt?

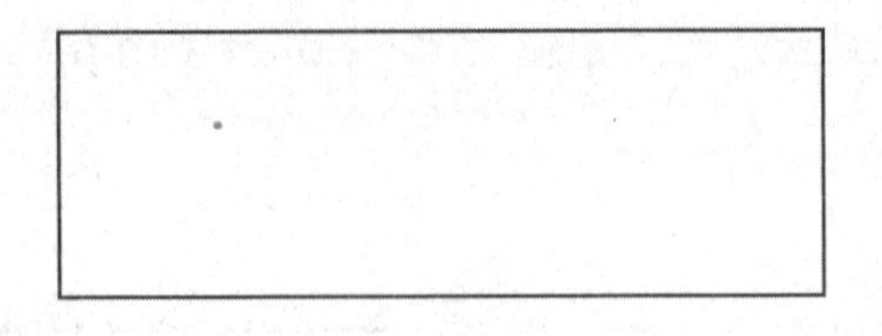

Name ______________________________ Date ______________

Solve the problems. Write your answers to show how many **tens** and **ones**.

[1 | 2] - 5 = 7
10 - 5 = 5
5 + 2 = 7

1. [1 | 7] - 8 = ____

____ - ____ = ____

____ + ____ = ____

2. [1 | 6] - 7 = ____

____ - ____ = ____

____ + ____ = ____

Solve. Write the two number sentences for each step to show how you take from **ten**. Remember to put a box around your solution and write a statement.

3. Yvette counted 12 kids at the park. She counted 3 on the playground and the rest playing in the sand. How many kids did she count playing in the sand?

---

4. Eli read some science magazines. Then, he read 9 sports magazines. If he read 18 magazines altogether, how many science magazines did Eli read?

____ - ____ = ____

____ + ____ = ____

5. On Monday, Paulina checked out 6 whale books and some turtle books from the library. If she checked out 13 books in all, how many turtle books did Paulina check out?

___ - ___ = ___

___ + ___ = ___

6. Some children are at the park playing soccer. Seven are wearing white shirts. If there are 14 children playing soccer in all, how many children are not wearing white shirts?

___ - ___ = ___

___ + ___ = ___

7. Dante has 9 stuffed animals in his room. The rest of his stuffed animals are in the TV room. Dante has 15 stuffed animals. How many of Dante's stuffed animals are in the TV room?

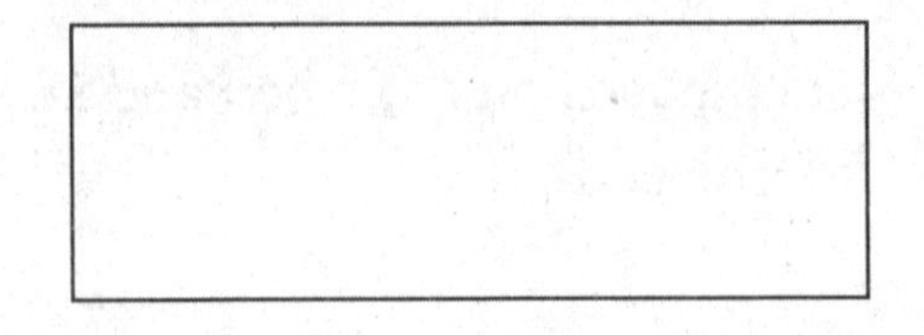

___ - ___ = ___

___ + ___ = ___